a

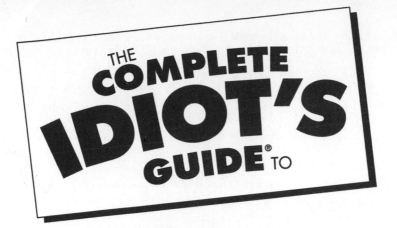

THE COMPLETE IDIOT'S GUIDE® TO

Precalculus

by W. Michael Kelley

A member of Penguin Group (USA) Inc.

For my twin baby girls, Erin and Sara. I prayed that God would send me a little girl to love, and He sent me two. Sure, you both had colic, but nobody ever said God didn't have a sense of humor.

ALPHA BOOKS

Published by the Penguin Group

Penguin Group (USA) Inc., 375 Hudson Street, New York, New York 10014, U.S.A.

Penguin Group (Canada), 10 Alcorn Avenue, Toronto, Ontario, Canada M4V 3B2 (a division of Pearson Penguin Canada Inc.)

Penguin Books Ltd., 80 Strand, London WC2R 0RL, England

Penguin Ireland, 25 St Stephen's Green, Dublin 2, Ireland (a division of Penguin Books Ltd.)

Penguin Group (Australia), 250 Camberwell Road, Camberwell, Victoria 3124, Australia (a division of Pearson Australia Group Pty Ltd.)

Penguin Books India Pvt Ltd., 11 Community Centre, Panchsheel Park, New Delhi—110 017, India

Penguin Group (NZ), cnr Airborne and Rosedale Roads, Albany, Auckland 1310, New Zealand (a division of Pearson New Zealand Ltd.)

Penguin Books (South Africa) (Pty) Ltd., 24 Sturdee Avenue, Rosebank, Johannesburg 2196, South Africa

Penguin Books Ltd., Registered Offices: 80 Strand, London WC2R 0RL, England

Copyright © 2005 by W. Michael Kelley

International Standard Book Number: 1-59257-301-0
Library of Congress Catalog Card Number: 2005922483

07 06 05   8 7 6 5 4 3 2 1

Interpretation of the printing code: The rightmost number of the first series of numbers is the year of the book's printing; the rightmost number of the second series of numbers is the number of the book's printing. For example, a printing code of 05-1 shows that the first printing occurred in 2005.

Printed in the United States of America

Note: This publication contains the opinions and ideas of its author. It is intended to provide helpful and informative material on the subject matter covered. It is sold with the understanding that the author and publisher are not engaged in rendering professional services in the book. If the reader requires personal assistance or advice, a competent professional should be consulted.

The author and publisher specifically disclaim any responsibility for any liability, loss, or risk, personal or otherwise, which is incurred as a consequence, directly or indirectly, of the use and application of any of the contents of this book.

Most Alpha books are available at special quantity discounts for bulk purchases for sales promotions, premiums, fundraising, or educational use. Special books, or book excerpts, can also be created to fit specific needs.

For details, write: Special Markets, Alpha Books, 375 Hudson Street, New York, NY 10014.

Publisher: *Marie Butler-Knight*
Product Manager: *Phil Kitchel*
Senior Managing Editor: *Jennifer Bowles*
Senior Acquisitions Editor: *Mike Sanders*
Development Editor: *Nancy D. Lewis*
Senior Production Editor: *Billy Fields*

Copy Editor: *Krista Hansing*
Cartoonist: *Shannon Wheeler*
Cover/Book Designer: *Trina Wurst*
Indexer: *Tonya Heard*
Layout: *Becky Harmon*
Proofreading: *Donna Martin*

Contents at a Glance

Contents

Introduction

I've never been a fan of "pre" anything. The entire purpose of tossing "pre" out in front of a word is to say, "Hey, eventually what you're doing will be important, but right now, not so much." Have you ever been subjected to a "pretest" at school? It basically looks and feels like the test you'll take at the end of a unit, but here's the catch: *You have to take it before the teacher actually teaches you anything!*

Supposedly, pretests help teachers "adjust their lesson plans to address the specific strengths and weaknesses of a class," but you and I both know that's not true. I was a teacher for a long time, and I'm not sure I ever saw anyone change their instruction as a result of a pretest, no matter what they say. The real reason you're given a pretest is to invoke this one thought deep in your subconscious mind: "Oh no! How am I ever going to learn any of this? I feel so overwhelmingly stupid, whatever will I do? Woe is me!"

Right when you start to panic and wonder if your mom was right when she tried to talk you out of taking the class ("The world needs its ditch-diggers, too," she'd say in a way that was really sweet and also really humiliating), the teacher steps in and says, "Don't worry—I'm here to save you from your ignorance. Open up your brains, and get ready to soak it all in." It's a tricky game of good cop/bad cop, where the pretest is the bad cop who wants you to fail, and the teacher is the good cop, who wants nothing more than for you to succeed (except perhaps for you not to slash her tires when she gives you detention).

So a "pretest" is not a test at all—it's just something you have to do before you're allowed to do something better and more fulfilling. Almost all "pre" words are like that. Olympics events have "prequalifying competitions," whose only purpose is to allow you to *get* to the Olympics, no matter how well you do. "Preschool" really isn't school so much as it is "learning how to go to school." Movie "previews" are only little clips from a film to get you excited about its release, which is often months, if not years, away!

"Pre" things are aggravating, often have no use in and of themselves, and (perhaps most frightening of all) are actually *commitments* more than anything else. When you take precalculus, there's an implicit assumption that you're looking to take calculus down the road, and it's going to be so hard that you have to start getting ready *now*.

Fortunately, precalculus is not your typical "pre" word. You're not just gearing yourself up to fight some ravenous calculus beast by learning skills that are, by themselves, useless. Even though no two precalculus courses are exactly the same, the majority of them feature advanced algebra, trigonometry, conic sections, and matrices—all tools that would be worth learning even if calculus didn't exist.

I suppose that's the one thing I want you to remember most as you grind your way through the topics in this book. The things you're learning are not pointless and are not here just to scare you into studying harder for a course that looms in the future like a blind date with awkward social skills. These topics are useful *right away*, for reasons that will make sense to you *right now*. So throw away your *pre*conceived notions, your *pre*eminently destructive fears, and your *pre*arranged marriages (although the last one is really none of my business), and embrace precalculus. (Or at least buy this book and embrace *it*, in case the going gets rough.)

How This Book Is Organized

This book is presented in five sections:

In **Part 1, "A Preponderance of Prioritized Precalculus Prerequisites,"** you'll make sure you're up to speed on the major algebra skills required in precalculus. Think of it as a brief "*pre*precalculus" course. Among the things you'll review are algebraic properties, exponential rules, polynomials, factoring, equations, inequalities, complicated fractions, and functions. You'll probably recognize a lot of the stuff in here, but you're bound to learn something new and to get a more thorough understanding of the things you were already somewhat familiar with.

Part 2, "Nonlinear Equations and Functions," broadens your horizons a little bit, taking the things you know about functions and equations and upping the ante. You'll start by solving quadratic equations (using three different techniques) and graduate to solving equations of higher degrees, like cubics and quartics. Then it's on to logarithmic and exponential functions, which are like nothing you've ever seen before. Not only are these graphs not simple lines, but they have minds all their own; that makes them a perfect introduction to the crazy graphs yet to come.

In **Part 3, "Trigonometry,"** you'll bask in the familiar sunshine of angles, triangles, circles, and other geometric memories for a while, and then things will start getting weird. All of a sudden, you'll be graphing repetitive things called periodic functions and manipulating complex strings of trig functions to try to prove that the identities they make up are actually true. By the end of the chapter, though, the hard work will pay off and you'll be measuring triangle components and calculating triangular areas that you'd never have dreamt possible back in the old geometry days. (If, that is, you often dream about such things, which I don't recommend.)

In **Part 4, "Conic Sections,"** words like *focus, vertex, center, radius,* and *directrix* will start popping up everywhere, not to mention all sort of axes—meaning more than one axis, not the "I am a homicidal maniac who needs firewood" kind of axes—such as the axis of symmetry, major axis, minor axis, transverse axis, and conjugate axis. All of these pieces come together to form four shapes in the coordinate plane: parabolas, circles, ellipses, and hyperbolas, which (all together) are called the conic sections.

Finally, in **Part 5, "Matrices and a Mathematical Montage,"** you'll introduce yourself to (or reacquaint yourself with) the relatively young field of mathematical study called matrices. They're nothing more than rows and columns of numbers (like a checkerboard with digits instead of squares), but with unfathomable superpowers. Whereas a checkerboard's only uses are playing checkers and swatting siblings, matrices can do all sorts of things, especially when it comes to systems of equations. After you feel like you've mastered matrices (and all the topics in the preceding chapters), try your hand at the Final Exam in Chapter 19. It'll help you figure out what you really do and don't understand.

Things to Help You Out Along the Way

As a teacher, I constantly found myself going off on tangents—everything I mentioned reminded me of something else. These peripheral snippets are captured in this book as well. Here's a guide to the different sidebars you'll see peppering the pages that follow.

Talk the Talk

Algebra is chock-full of crazy- and nerdy-sounding words and phrases. To become King or Queen Math Nerd, you'll have to know what they mean!

Critical Point

These notes, tips, and thoughts will assist, teach, and entertain. They add a little something to the topic at hand, whether it be some sound advice, a bit of wisdom, or just something to lighten the mood a bit.

Kelley's Cautions

Although I will warn you about common pitfalls and dangers as I explain precalculus topics, the dangers in these boxes deserve special attention. Think of these as skulls and crossbones painted on little signs that stand along your path. Heeding these cautions can save you hours of frustration.

> ## How'd You Do That?
>
> All too often, algebraic formulas appear like magic, or you just do something because your teacher told you to. If you've ever wondered "Why does that work?", "Where did that come from?", or "How did that happen?", this is where you'll find the answer.

> ## You've Got Problems
>
> Math is not a spectator sport! Once I explain how to do a problem, you should try an example on your own. These practice items will be very similar to those I walk you through in the chapters, but now it's your turn to shine. You'll find all the answers, explained step by step, in Appendix A.

Acknowledgments

There are two distinct parts of me, Book Mike and Regular Mike, and both require constant attention, help, support, love, and regular poker games to survive. There is a small group of folks who not only allow me to write, but actually encourage it (so blame them). Special thanks to Jessica Faust (my agent), Mike Sanders (at Alpha Books), Nancy Lewis (my longsuffering and fantastic editor), Sue Strickland (one of those rare college professors who really means it when she says she cares about you), and my first real friends in the publishing biz (who still take my calls, believe it or not): Roxane Cerda and David Herzog.

My support structure of family and friends is just as reliable, if not so obsessed with deadlines. To Dave (my brother); Carol (my mom); Matt "The Prophet" Halnon, Chris "The Cobra" Sarampote, and Rob "Nickels" Halstead (my closest friends and Texas Hold'Em nemeses); Lisa (my wife, my best friend, and my light in the darkness); Nicholas (my son and my proudest accomplishment): I have no words to express my thanks for the things you do for me every day without even being asked.

Finally, thanks to Erin and Sara, two little girls I love with all my heart. Thank you for adding immeasurable joy to my life, and for not being born until I got this whole book written (even though you did cut it pretty close).

Special Thanks to the Technical Reviewer

The Complete Idiot's Guide to Precalculus was reviewed by an expert who double-checked the accuracy of what you'll learn here, to help us ensure that this book gives you everything you need to know about getting ready for calculus. Special thanks are extended to Susan Strickland, who provided the same service for *The Complete Idiot's Guide to Algebra* and *The Complete Idiot's Guide to Calculus*.

Susan Strickland received a Bachelor's degree in mathematics from St. Mary's College of Maryland in 1979 and a Master's degree in mathematics from Lehigh University in 1982. She took graduate courses in mathematics and mathematics education at The American University in Washington, D.C., from 1989 through 1991. She was an assistant professor of mathematics and supervised student teachers in secondary mathematics at St. Mary's College of Maryland from 1983 through 2001. During that time, she had the pleasure of teaching Michael Kelley and supervising his student teaching experience. Since 2001, she has been a professor of mathematics at the College of Southern Maryland and is now involved with teaching math to future elementary school teachers. Her interests include teaching mathematics to "math phobics," training new math teachers, and solving math games and puzzles (she can really solve the Rubik's Cube).

Trademarks

Part 1

A Preponderance of Prioritized Precalculus Prerequisites

You've got to be pretty good at algebra before you attempt this precalculus course. That's why, in this part, I help you review the algebra skills you'll need later. Of course, I have to move a bit quickly because most of the chapters should be review for you. If you need extra help on any of these topics, might I suggest my prequel, *The Complete Idiot's Guide to Algebra*? It's one prequel (unlike *Star Wars Episode One: The Phantom Menace*) that isn't painful to sit through.

Reviewing Numbers and Arithmetic

In This Chapter

- ◆ Classifying numbers according to their properties
- ◆ The underlying rules of algebra
- ◆ The laws of exponents
- ◆ Simplifying radical expressions

For a lot of reasons, I will never skydive. For one thing, I have a paralyzing fear of heights. My longstanding love affair with the ground is well established, but if I leave it for just a moment to climb somewhere high (usually against my will), I still have no desire to see it rushing at me for an impromptu reunion due to a misstep or an unlucky fall.

However, if I ever decided to give freefalling from airplanes a try, there's one thing I'd be sure to do: I'd spend a lot of time getting ready. I'd want to have every skill mastered, no matter how simple, so that halfway through the jump, I wouldn't suddenly realize that I didn't possess an important skill (such as how to avoid hitting a barn or how to swallow the fewest number of bugs while my mouth is stretched in a scream of horror).

Learning precalculus is not nearly as physically strenuous or terrifying as skydiving, but the course lasts a whole lot longer than a single jump does. If you don't have all of its mathematical prerequisites mastered, you might suddenly find yourself metaphorically tangled in your parachute or, even worse, plummeting to your educational doom, with only the sounds of incomprehensible variables and equations whizzing by at impossible speeds.

In this chapter, I'll review the most basic algebra skills and vocabulary terms you'll need to know once you get into the meatier precalculus topics beginning in Part Two. Even if you think you've got a good grip on algebra, you should review the chapter because it would be tragic to find out later that the bag you thought contained your parachute was actually empty.

Number Classifications

During your life, you've learned quite a bit about numbers, but you'll need to know even more when it comes to *describing* numbers. Throughout the book, I'll refer to major number classifications, such as the rational numbers or the integers, so it's important that you know what I'm talking about. Here they are listed in order, from the smallest number group to the largest:

- **Natural numbers.** These are the numbers you used when you played Hide and Seek. While your playmates scampered around, disappearing into shrubs, you'd stand with your eyes squeezed shut, progressing through the natural numbers: "1, 2, 3, 4, 5, 6, …." In fact, for just this reason, this group is also known as the *counting numbers*. Note that the smallest natural number is 1, and the group contains no fractional, decimal, or negative numbers.

> **CAUTION**
>
> **Kelley's Cautions**
>
> Technically, natural numbers can contain inconsequential fractions and decimals. For instance, $\frac{8}{1}$ and 8.00 are natural numbers because they can be rewritten as 8, a number that needs no fraction bar or decimal point.

- **Whole numbers.** If you insert one more member (the number 0) into the set of natural numbers, you get the whole numbers: 0, 1, 2, 3, 4, 5, 6, …. Basically, a whole number is any non-negative number that doesn't contain a fraction or a decimal.

- **Integers.** To generate the complete list of integers, tack on –1, –2, –3, –4, … (the opposites of every natural number) to the set of whole numbers.

♦ **Rational numbers.** Any number that can be expressed as a fraction $\frac{a}{b}$ (where a and b are both integers) is automatically a rational number. Therefore, $\frac{2}{3}$, $8\frac{1}{2}$, and $-\frac{7}{5}$ are all rational numbers. Here's something you might not know: If a decimal either infinitely repeats (as in 5.172172172172172…) or terminates in a fixed number of decimal places (as in 3.7 or –14.98132), that's proof enough that the number is actually a fraction in disguise and, therefore, a rational number.

♦ **Irrational number.** Any number that does not meet the requirements to fit in with the rational numbers gets grouped with the irrational numbers. Therefore, to be irrational, the number must be a nonrepeating, nonterminating decimal (like π = 3.14159265358979323846264338832795…, which never follows any repetitive pattern, no matter how many decimal places you include), and it must be impossible to express in fraction form.

♦ **Real number.** What do you get when you combine all of the rational and irrational numbers? A dangerous, volatile chemical weapon? An uncomfortable school dance where each group of numbers stands on separate sides of the gym and won't talk to each other? No, you get the real numbers. Just like men and women, as a group, are referred to as "humans," the rationals and irrationals together are known as the "reals."

♦ **Complex numbers.** A complex number has the form $a + bi$, where a and b are both real numbers and i has the bizarre value $i = \sqrt{-1}$. Since you can't actually take the square root of a negative number, i is described as imaginary. Therefore, in the complex number $a + bi$, a is called the real part and bi is the imaginary part.

You'll work with complex numbers only a little bit in precalculus, but you do need to know how to perform basic arithmetic operations on them; I'll go through this at the end of Chapter 3.

Critical Point _____

Any integer is automatically a rational number as well because every integer can be expressed as the quotient of itself and 1. For instance, $13 = \frac{13}{1}$, so 13 is definitely rational. And any real number is automatically a complex number. Because you can rewrite −5 as −5 + 0i, −5 is complex even though its imaginary part, 0i, technically equals 0.

Keep in mind that most numbers can be classified in many different ways and don't belong to only one group. Just as you could classify a whale as a mammal and as "something too big to fit comfortably into a minivan," the number 7 can be classified as a natural number, a whole number, an integer, a rational number, a real number, a complex number, and a really gross/awesome movie starring Brad Pitt and Morgan Freeman.

Algebraic Properties

A *property* is a mathematical fact that is so basic and fundamental that it is accepted as truth (even though you can't rigorously prove it true). Math, like any logical or belief system, has certain foundational truths that, even though they are "obviously right," cannot be empirically proven. The most commonly referenced algebraic properties are listed here:

Talk the Talk

The most basic, unprovable rules that form the foundation of algebra are called the algebraic **properties** (or axioms).

Kelley's Cautions

Remember, the associative and commutative properties work only with addition and multiplication. If you regroup or reorder subtraction or multiplication problems, you most likely change the result. For instance, $10 \div 2$ and $2 \div 10$ are definitely unequal.

- **Associative property (of addition and multiplication).** If you're given a sum (a list of real numbers all added together) or a product (numbers all multiplied together), you can group those numbers however you like (by placing parentheses anywhere), and it will not change the result. Notice that the expressions that follow have the same value, even though the placement of the parentheses is different.

$$(3+4+1)+6+(7+11)$$
$$=8+6+18$$
$$=32$$

$$3+(4+1+6+7)+11$$
$$=3+18+11$$
$$=32$$

- **Commutative property (of addition and multiplication).** Given a sum or product, you can alter the order of the numbers involved, but it won't change the value of the sum or product.

$$4 \cdot 5 \cdot (-3) \cdot 2 \qquad\qquad 2 \cdot 5 \cdot 4 \cdot (-3)$$
$$= 20 \cdot (-3) \cdot 2 \qquad\qquad = 10 \cdot 4 \cdot (-3)$$
$$= -60 \cdot 2 \qquad\qquad\quad = 40 \cdot (-3)$$
$$= -120 \qquad\qquad\quad = -120$$

- ◆ **Identity properties (of addition and multiplication).** If you add 0 to any real number ($a + 0$) or multiply a real number by 1 ($a \cdot 1$), you get the original number back, unchanged (a). This is not earth-shattering news, I know. However, you should remember that, as a result, 0 is called the *additive identity* and 1 is known as the *multiplicative identity*. (When applied, these two numbers do not change the *identity* of other numbers or variables.)

- ◆ **Inverse properties.** Every real number b has an *opposite*, written $-b$, such that the sum of a number and its opposite equals 0: $b + (-b) = 0$. Furthermore, every nonzero real number b also has a *reciprocal*, defined as $\frac{1}{b}$, and when you multiply a number by its reciprocal, you get 1: $b \cdot \frac{1}{b} = \frac{b}{b} = 1$.

Talk the Talk

The identities of addition and multiplication are (respectively) 0 and 1. Notice that when an inverse property is applied to a number, the result is always one of these identities. (When you add **opposites**, you get the **additive identity**, and when you multiply **reciprocals**, you get the **multiplicative identity**.

- ◆ **Symmetric property.** You can reverse the sides of an equation without affecting its solution. In other words, if $x = y$, then $y = x$. The symmetric property is most often used when you're trying to solve an equation and you want to isolate the variable on a particular side, usually the left side of the equation. Even though a final solution of $-4 = x$ is acceptable, most people prefer the more aesthetic-looking solution $x = -4$.

- ◆ **Transitive property.** If $a = b$ and $b = c$, then $a = c$. Think of the transitive property this way: If you are exactly as tall as I am, and I am exactly as tall as William Shatner (which is true—we're both 5 feet, 11 inches), then you must be exactly as tall as William Shatner.

♦ **Distributive property.** The expression $a(b + c)$ is equal to $ab + ac$. In other words, you can "distribute" the a through the parentheses, multiplying it by each enclosed term. You can also distribute over a subtraction problem or a combination of addition and subtraction, like this: $a(b + c - d) = ab + ac - ad$.

♦ **Substitution property.** If two expressions are equal, you can replace one with the other. For instance, if you are given that $2x - 3y = 9$ and you know that $x = 5y$, you can substitute $5y$ for x in the equation: $2(5y) - 3y = 9$.

♦ **Monopoly properties.** If you ever play Monopoly against algebra, keep in mind that it always tries to buy up the red properties: Illinois, Indiana, and Kentucky Avenues. That way, when you just miss Free Parking, insult is added to injury when you have to pay rent. Diabolical.

Even though you may be tempted to substitute your own, more understandable names for the properties (such as referring to the symmetric property as the "flip-flop property"), learn the real names and stick to them. That way, it's easier to communicate your thoughts and justifications because even though you may know what you mean by "You know, that one that says you can move stuff?", other people may be a bit confused.

You've Got Problems

Problem 2: Identify the properties that justify the following statements:

   a. $4 + 1 = 9 - 4$ is equivalent to $9 - 4 = 4 + 1$

   b. $(3)(-7)(8) = (8)(-7)(3)$

   c. $4(x - 2) = 4x - 8$

   d. $\left(-\dfrac{3}{2}\right)\left(-\dfrac{2}{3}\right) = 1$

Exponential Rules

You can indicate repeated multiplication of a number, variable, or quantity by using *exponents*. In the expression x^7 (read "x to the seventh power"), x is the *base* and 7 is the *exponent*, or *power*. The notation x^7 means the same thing as $x \cdot x \cdot x \cdot x \cdot x \cdot x \cdot x$. In addition, you can rewrite the expression $5y \cdot 5y \cdot 5y \cdot 5y \cdot 5y \cdot 5y$ as $(5y)^6$, where $5y$ is the *base* raised to a *power* of 6 (the total number of times the base appears multiplied times itself).

Even though exponential notation is technically only a quick way to rewrite a problem, you can easily work with exponential expressions without having to rewrite them the long way first; just follow these rules:

- $x^a x^b = x^{a+b}$: If two exponential expressions (with the same base) are multiplied, the product is the shared base raised to the *sum* of the powers: $w^3 \cdot w^9 = w^{3+9} = w^{12}$. Even though the expression $y^2 \cdot y^3$ is a product, the answer is y^5, *not* y^6; you must add the exponents, not multiply them. Remember that exponents are just short-hand for repeated multiplication, so $y^2 \cdot y^3$ actually stands for $(y \cdot y)(y \cdot y \cdot y)$. According to the associative property, you can regroup all those y's together to get $y \cdot y \cdot y \cdot y \cdot y = y^5$.

- $\dfrac{x^a}{x^b} = x^{a-b}$: If two exponential expressions (again, they must have the same base) are divided, the result is the common base raised to the difference of the powers: $\dfrac{b^{12}}{b^7} = b^{12-7} = b^5$. Note that you should always subtract the denominator power from the numerator power. The previous rule stated that multiplying exponential expressions implied adding the powers, so it makes sense that *dividing* exponential expressions would imply *subtracting* powers—each is the reverse of the other.

- $(x^a)^b = x^{a \cdot b}$: If an exponential expression is, itself, raised to an exponent, multiply the exponents together and keep the shared base: $(k^3)^4 = k^{3 \cdot 4} = k^{12}$. Another option is to expand the exponential expression: $(k^3)^4 = (k^3)(k^3)(k^3)(k^3) = k^{3+3+3+3} = k^{12}$.

- $(xy)^a = x^a y^a$ and $\left(\dfrac{x}{y}\right)^a = \dfrac{x^a}{y^a}$: If an entire product or quotient is raised to an exponent, each individual piece of that expression gets raised to the exponent: $(x^2 y)^3 = (x^2)^3 \cdot (y)^3 = x^6 y^3$.

- $x^1 = x$ and $x^0 = 1$: Any number raised to the first power is equal to the number itself, so $13^1 = 13$ and $(-4)^1 = -4$. (Because this is true, there's really no need to write an exponent of 1.) Additionally, if you raise any number (except 0) to the zero power, you get 1: $13^0 = 1$ and $(-4)^0 = 1$. You won't deal with the value of 0^0 until calculus.

Kelley's Cautions

If a sum or difference is raised to an exponent, you *cannot* raise each piece to that exponent, as in $(x + y)^2 \neq x^2 + y^2$ and $(a - b)^3 \neq a^3 - b^3$. I review this type of expansion problem in Chapter 3.

◆ $x^{-a} = \dfrac{1}{x^a}$ and $\dfrac{1}{x^{-a}} = x^a$: A negative exponent indicates that the number it's attached to is in the wrong part of the fraction; if you move that number across the fraction bar (from the numerator to the denominator or vice versa), that solves the problem and the negative exponent becomes positive again.

Critical Point

If an entire fraction is raised to a negative power, you can take the reciprocal of the fraction and make the exponent positive:

$$\left(\frac{x}{y^2}\right)^{-2} = \left(\frac{y^2}{x}\right)^2 = \frac{y^4}{x^2}$$

Take, for example, the fraction $\dfrac{x^{-4}y^2}{z^{-1}}$. Since the x and z have negative powers, move them each across the fraction bar to get $\dfrac{y^2 z}{x^4}$. Why do this? Most math teachers consider a solution containing negative exponents to be unsimplified, so you've got to ditch those negative powers in your final answer.

Example 1: Simplify the expression $(m^2 p)^3 \cdot (mp^2)^{-2}$.

Solution: Note that each exponent is raised to an exponent, so multiply the powers:

$$(m^{2(3)}p^{1(3)})(m^{1(-2)}p^{2(-2)}) = m^6 p^3 \cdot m^{-2} p^{-4}$$

According to the commutative property, you can rearrange the order to pair up like bases and then add the exponents connected to the like bases:

$$m^6 m^{-2} \cdot p^3 p^{-4} = m^{6-2} p^{3-4} = m^4 p^{-1}$$

Eliminate the negative exponent by moving p to the denominator.

$$\frac{m^4}{p}$$

You've Got Problems

Problem 3: Simplify the expression $\left(\dfrac{x^3 y^5}{x^7 y^2}\right)^{-2}$.

Radical Expressions

So far, I have discussed only exponents that are integers, but it is not only possible to have rational exponents, it is also quite common. However, these fractional powers skulk about in disguise, dressed up as radical expressions. (Unfortunately, they are not nearly as exciting as they sound. You probably won't refer to them as "radical" in the descriptive sense unless you are a die-hard skate punk excited about algebra, and I'm pretty sure those are few and far between.)

Fractional Powers

A *radical expression* looks like this: $\sqrt[b]{a}$ (read "the *b*th root of *a*"). Be sure to note that *a* could be any real number, but *b* must be a positive integer. The number *a* inside the radical sign is called the *radicand*, and the teeny weeny *b* nestled in the check mark part of the radical sign is called the *index*.

As I just mentioned, a radical expression is really the result of a fractional power. Specifically, they are defined like this: $a^{1/b} = \sqrt[b]{a}$. The denominator of the fractional power turns into the index of the radical expression:

$$x^{1/4} = \sqrt[4]{x} \qquad 7^{1/2} = \sqrt{7} \qquad (5w)^{1/3} = \sqrt[3]{5w}$$

A fractional power, however, does not always have a numerator of 1, like the previous examples do. When translating a rational exponent into a radical expression, the numerator turns into either the power of the radicand or the exponent of the *entire radical expression*. For example, $x^{2/3}$ can be rewritten as either $\sqrt[3]{x^2}$ or $\left(\sqrt[3]{x}\right)^2$. (In either case, the denominator of the rational exponent always turns into the index of the radical expression.)

Critical Point _____

If a radical expression does not have an index (like $\sqrt{7}$), it is assumed to be 2. Such expressions are called *square roots*. (Expressions with an index of 3 are called *cube roots*.)

<hr>

You've Got Problems

Problem 4: Rewrite the following as radical expressions.
   a. $y^{1/2}$
   b. $b^{3/2}$

Simplifying Radical Expressions

If algebra taught you anything, it was that (for some inexplicable reason) simplifying things was the reason you were alive. Fraction not fully reduced? Attack! Is that a polynomial containing like terms that aren't combined properly? Move in and neutralize the targets immediately! Here are a few examples to help you review simplifying radical expressions so that you're ready to mobilize at a moment's notice and strike down those unholy, unsimplified monsters.

Example 2: Simplify the radical expressions.

a. $\sqrt{32x^3}$

Since the index of this expression is 2, you want to rewrite 32 and x^3 so that they contain as many exponents of 2 as possible. (Factor both using the largest possible perfect squares.)

$$\sqrt{16 \cdot 2 \cdot x^2 \cdot x} = \sqrt{4^2 \cdot 2 \cdot x^2 \cdot x}$$

Since the powers of 4^2 and x^2 match the index of the radical, they get pulled out in front of the radical sign (without their exponents), but the 2 and x in the radicand must stay behind: $4|x|\sqrt{2x}$. Check the nearby caution sidebar if you're wondering where those absolute value bars came from.

Kelley's Cautions

If a variable's even power n equals the index of the radical $\left(\sqrt[n]{x^n}\right)$, you must write absolute value signs when you simplify.

b. $8^{4/3}$

Begin by rewriting the rational exponent as a radical expression: $\left(\sqrt[3]{8}\right)^4$. (You could also rewrite it as $\sqrt[3]{8^4}$, but 8^4 is a pretty big number and will be much harder to simplify.) Since $\sqrt[3]{8} = 2$, substitute 2 into the expression, and you'll have $(2)^4$, which equals 16.

c. $\sqrt[3]{16y} - \sqrt[3]{54y}$

Don't simply add the radicals together. You can combine the coefficients of radicals only if the radicands match exactly. (The indices have to match, too, but in this case, they're both 3, so that's okay.) Try to simplify the radicals using as many perfect cubes as possible so that you can pull those exponents of 3 out of the radical sign.

$$\sqrt[3]{2^3 \cdot 2 \cdot y} - \sqrt[3]{3^3 \cdot 2 \cdot y} = 2\sqrt[3]{2y} - 3\sqrt[3]{2y}$$

Now that both radicals in the expression have matching indices (3) and matching radicands (2y), you can combine their coefficients (2 − 3 = −1) to get $-1\sqrt[3]{2y}$, or just $-\sqrt[3]{2y}$.

> **You've Got Problems**
>
> Problem 5: Simplify the following expressions.
> a. $16^{5/2}$
> b. $\sqrt{28x^2} + \sqrt{63x^2} - \sqrt{7x^2}$

The Least You Need to Know

♦ The basic algebraic number classifications are (from smallest to largest): natural, whole, integer, rational, irrational, real, and complex.

♦ Algebraic properties, while technically not provable, are nonetheless accepted as fact.

♦ You must adhere to specific laws when simplifying exponential expressions.

♦ Radical expressions are generated by rational exponents.

2

Equations and Inequalities

In This Chapter

- ◆ Solving equations in one variable
- ◆ Designing and graphing linear equations
- ◆ Finding solutions for linear inequalities
- ◆ Applying interval notation to inequalities
- ◆ Graphing one- and two-dimensional inequalities

What fun would math be if it dealt with only plain old numbers? Once you learned how to add, subtract, multiply, and divide, you'd know just about everything there was to know. Ultimately, you would be unsatisfied because we, as humans, crave a little mystery in life—we desire the unknown. In math, the unknowns are really easy to identify—they're little italicized letters you know and love as variables.

Sure, variables can sometimes cause more trouble than they're worth. There will be moments when they defy you and, despite all your best efforts, you can't figure out what value they're hiding, but let's face it—that's one of the reasons we love them. They're sort of naughty, always trying to elude you, but also always pleased when you figure them out, despite all their mysterious airs. Even when you can finally pin one down and proclaim

"Ah ha, $x = -2$, gotcha!" the variable grins slyly, with a look that says "Maybe this time, but we'll see if you can track me down on your next assignment."

In this chapter, you'll review basic one- and two-variable equations and inequalities, where you and variables first met. I've invited you both here, to a neutral place, without telling either of you, so that you can fall in love all over again (sort of like *The Parent Trap*, with a math-flavored twist.)

Solving Equations in One Variable

Equation-solving strategy all boils down to one principle: Whatever you do to one side of the equation, you should do to the other as well, or the delicate balance of equality is ruined. Think of the sides of an equation as twin children: If you don't treat them both equally, your whole family will soon be out of whack.

Mathematically speaking, treating the sides of an equation the same means always adding, subtracting, multiplying, or dividing by the same number on both sides, with one goal in mind: to isolate the variable so it sits all alone on one side of the equation and the solution sits on the other.

Example 1: Solve the equation $-\frac{1}{2}(3x - 8) = 25$.

Solution: Start by distributing $-\frac{1}{2}$ through the parentheses:

$$-\frac{3}{2}x + 4 = 25$$

Isolate the term containing the variable by subtracting 4 from both sides of the equation.

$$
\begin{array}{rccc}
-\frac{3}{2}x & + & 4 & = 25 \\
 & & -4 & -4 \\
\hline
-\frac{3}{2}x & & & = 21
\end{array}
$$

Critical Point _____

Remember, to multiply a fraction and an integer, you can rewrite the integer as a fraction with a denominator of 1. In Example 1,

$$-\frac{1}{2} \cdot 3 = -\frac{1}{2} \cdot \frac{3}{1} = -\frac{3}{2}.$$

To isolate x (and, therefore, solve the equation), multiply both sides by the reciprocal of x's coefficient:

$$-\frac{3}{2}x = 21$$

$$-\frac{2}{3}\left(-\frac{3}{2}x\right) = -\frac{2}{3}\left(\frac{21}{1}\right)$$

$$\frac{6}{6}x = -\frac{42}{3}$$

$$x = -14$$

To check the answer, you can plug it back into the original equation for x and see what happens:

$$-\frac{1}{2}(3(-14)-8)$$

$$= -\frac{1}{2}(-42-8)$$

$$= -\frac{1}{2}(-50)$$

$$= 25$$

Since you end up with a true statement at the end, the solution of $x = -14$ must have been correct.

You've Got Problems

Problem 1: Solve the equation $-5\left(\frac{2}{3}x+6\right) = \frac{7}{3}x+4$.

Writing Equations of Lines

You can construct the equation of any line given only three important elements:

- The slope of the line, m
- A point on the line (a,b)
- A bundt pan full of live mice

I honestly don't know why the final ingredient is necessary, but who am I to argue with famed mathematician René Descartes, renowned founder of analytical geometry and lover of rodents? Truth be told, you can actually construct the equation using only the first two things (but trust me, the ambience just won't be the same).

Talk the Talk

If a line has slope m and contains the point (a,b), then it has the equation $y - b = m(x - a)$, called the **point-slope formula**.

To create the equation of a line, simply plug the slope m and point (a,b) into the *point-slope formula*. (I'll bet you'll never guess how they came up with that name.)

$$y - b = m(x - a)$$

Some textbooks use the formula $y - y_1 = m(x - x_1)$ to describe the point-slope formula; this is because they use (x_1,y_1) to denote the point instead of (a,b), but either formula will give you the same answer.

Once you plug in the correct slope and point values, you may be asked to write the equation in one of two specific forms:

♦ **Standard form.** The x and y terms should be on the left side of the equation, and the number should be on the right. Furthermore, neither the x nor the y coefficients can be a fraction, and the x coefficient must be positive. Mathematically speaking, an equation is in standard form if it looks like $Ax + By = C$, where A and B are integers and $A > 0$. By the way, if a line is in standard form, its slope is equal to $-\dfrac{A}{B}$; that shortcut sometimes comes in very handy.

♦ **Slope-intercept form.** A linear equation is automatically in slope-intercept form once you isolate the y on the left side of the equation. This is often a valuable thing to do because the resulting x coefficient is the slope of the line, and the constant is the line's y-intercept. In other words, the equation $y = mx + b$ is in slope-intercept form because it is solved for y; the slope of that line is m and its y-intercept is b.

Kelley's Cautions

You'll be writing equations of lines for the rest of your mathematical career, so make sure you memorize the formulas for point-slope, slope-intercept, and standard forms. (And also be sure to keep a bundt cake pan with you at all times.)

If you're not specifically instructed by a problem to use one of those forms, I would default to slope-intercept form. Because your result from the point-slope formula is nearly solved for y already, it takes only a smidgen of effort.

Example 2: Write the equation of each line in the form indicated.

a. Write line k, which contains points $(3,-1)$ and $(2,9)$, in slope-intercept form.

Solution: In algebra, you learned that the slope connecting two points (x_1, y_1) and (x_2, y_2) was equal to $\dfrac{y_2 - y_1}{x_2 - x_1}$:

$$m = \frac{9 - (-1)}{2 - 3} = \frac{10}{-1} = -10$$

Now that you know the slope, use it and either of the given points in point-slope form. I'll use the point $(3,-1)$, but you'd get the same final answer using $(2,9)$.

$$y - b = m\left(x - a\right)$$
$$y - (-1) = -10\left(x - (3)\right)$$
$$y + 1 = -10x + 30$$

To write this in slope-intercept form, solve the equation for y (by subtracting 1 from both sides):

$$y = -10x + 29$$

b. Write line j in standard form if you know that j has a y-intercept of -5 and is perpendicular to the line whose equation is $2x - y = 5$.

Solution: The slope of $2x - y = 5$ is $-\dfrac{2}{-1} = 2$ (according to the shortcut I mentioned for standard form), so the slope of j must be the opposite reciprocal: $-\dfrac{1}{2}$. If j has a y-intercept of -5, then it must pass through the point $(0,-5)$. Use this point and slope in the (you guessed it) point-slope formula:

Critical Point

Don't forget this important algebra fact: Parallel lines have equal slopes, and perpendicular lines have slopes that are opposite reciprocals.

$$y - (-5) = -\frac{1}{2}\left(x - (0)\right)$$
$$y + 5 = -\frac{1}{2}x$$

The x and y terms must be on the left side of the equation in standard form (and the constant, 5, must move to the right).

$$\frac{1}{2}x + y = -5$$

The x coefficient is now positive, which meets another requirement of standard form, but it is not an integer (remember, both it and the y coefficient must be integers), so multiply both sides of the equation by 2 to kill off the fraction.

$$2\left(\frac{1}{2}x + y\right) = 2(-5)$$
$$x + 2y = -10$$

You've Got Problems

Problem 2: If the point (–7,–3) lies on line w, and w is parallel to line v (whose equation is $6x + y = 4$), write the equation of w in slope-intercept form.

Graphing Linear Equations

The graph of a line, like any graph on the coordinate plane, consists of all the points (x,y) that make the equation true. How many points actually make up the graph of a line? An infinite number (you know exactly what I'm talking about if you've ever waited in line to ride the Space Mountain roller coaster in Disney World).

Luckily, however, all of the points on a linear graph behave in a predictable fashion—they all stand neatly and orderly in a row, like ants returning to the hill from a fallen sandwich—so it's unnecessary to plot more than two points to get a graph that's just as precise as one containing 3,000 points. The method I describe in Example 3 (called the "intercept method" by some textbooks) is just one of many ways to graph a line, but I think it's the easiest.

CAUTION **Kelley's Cautions** _____

The intercept method doesn't work if there's only one variable in the equation, as in $x = 3$ or $y = -1$, but you really don't need it to. The graph of a line with the equation $x = a$ is just a vertical line on the coordinate plane at that a value. For instance, the graph of $x = -2$ is an infinitely long vertical line two units to the left of the origin.

Similarly, the graph of $y = b$ is a horizontal line at b. There's no need to do any real work to generate these graphs—they're automatic.

Example 3: Graph the linear equation $3x - 2y = 6$.

Solution: One at a time, I will substitute 0 for x and then y to get the intercept value for the other variable. Even though I'll start by setting $x = 0$, I could just as easily and just as correctly have started with $y = 0$; the order doesn't matter.

$$3(0) - 2y = 6$$
$$-2y = 6$$
$$y = -3$$

This line will cross the y-axis at the point $(0, -3)$. To find the x-intercept, go back to the original equation, this time substituting in $y = 0$:

$$3x - 2(0) = 6$$
$$3x = 6$$
$$x = 2$$

The line must hit the x-axis at the point $(2, 0)$. To generate the graph, simply plot the points $(0, -3)$ and $(2, 0)$ on the coordinate plane and connect them with a line, as in Figure 2.1. Remember, that line extends infinitely in either direction. It's not a segment that has those endpoints.

Figure 2.1

The graph of 3x − 2y = 6 is the line connecting and extending through its x-intercept (2,0) and its y-intercept (0,−3).

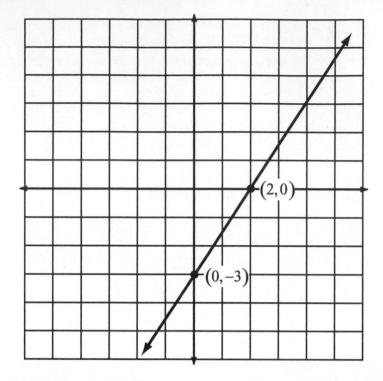

You've Got Problems

Problem 3: Graph the linear equation $x - 3y = -5$.

Solving Linear Inequalities

Inequalities are a lot like equations. They like the same kind of music (adult contemporary, oddly enough), enjoy the same Kevin Costner movies (none), and both secretly hope one day to be on *The Price Is Right*. In fact, you could think of equations and inequalities as fraternal twins, the Mary-Kate and Ashley Olsen of the mathematical world. However, fraternal twins do have very subtle differences, and it's those subtleties you need to understand.

Interval Notation

In precalculus, you use *interval notation* to express the answers to inequalities. It's not a new type of problem you have to master—just a new way to write answers to problems you can already solve. An *interval* is a piece of the number line, even though it just looks like a garden-variety pair of numbers inside parentheses, brackets, or a combination of the two. Those numbers are the solution's boundaries written in order, the lower boundary first and the upper boundary second.

You'll have to decide whether to use a bracket or parenthesis for each boundary of the interval separately, and here's the question you'll ask yourself to reach that decision: "Is this a finite, real number that is included in the rest of the group?" If the answer is yes, use a bracket. If the answer to any part of that question is no, use a parenthesis. Remember to investigate each boundary separately because including or excluding one doesn't imply that you'll do the same for the other.

Talk the Talk

Interval notation is a shorthand method for expressing inequality statements that consists of two boundary numbers surrounded by parentheses or brackets, depending upon whether each boundary is included in the solution interval.

Example 4: Rewrite the inequalities using interval notation.

a. $-1 < x < 5$

Solution: The lower and upper bounds couldn't be clearer; they're -1 and 5, respectively. However, -1 and 5 are not actually members of the group. The inequality specifically says x must be less than 5 (not less than *or equal to* 5) and greater than -1 (not greater than *or equal to* -1). Since the boundaries are excluded, use parentheses around each one for an answer of $(-1,5)$, an *open interval*.

b. $3 \leq y \leq 10$

Solution: This time, the boundaries are both included, indicating a *closed* interval, so use brackets: $[3,10]$.

c. $9 > p \geq -2$

This inequality is written with its upper boundary first. Remember, interval notation requires the lower boundary first, so -2 must come before 9. Also notice that -2 is included in the interval (since the sign next to it contains the clause "or equal to") and, therefore, deserves a bracket. The upper boundary, 9, gets only a parenthesis: $[-2,9)$.

Talk the Talk

An interval with two parentheses is called an **open interval**. If two brackets are present, the interval is said to be **closed**. However, if it contains one parenthesis and one bracket, it can be described as either **half-open** or **half-closed**.

d. $x > 3$

The lower boundary is clear here; you cannot go lower than 3. However, there is no finite upper boundary, so use ∞ (infinity). Note that you *must* use a parenthesis next to an infinite boundary, since one of the requirements for a bracket is that the number be finite: $(3, \infty)$.

e. $p \leq -9$

This time, the finite upper boundary is obvious (–9) and included in the interval. Now the lower boundary is infinite, so you write –∞: $(-\infty, -9]$. Notice that an infinite boundary is always negative when it's the lower boundary of an interval, and is always positive when it's the upper boundary.

You've Got Problems

Problem 4: Rewrite the following in interval notation.

      a. $y \geq -1$
      b. $9 < z \leq 13$
      c. $x \neq 4$ *(Hint: requires two intervals)*

Simple Inequalities Taste Like Chicken

If you can solve an equation, you can solve an inequality. They act, smell, feel, and taste almost the same. Just remember one thing: If you ever multiply or divide both sides of an inequality by a negative number, you must reverse the inequality sign. By "reverse," I mean change < into >, or change ≥ into ≤. The possibility of equality will not change (so < won't become ≥)—only the direction the inequality sign points will change.

Example 5: Solve the inequality and graph the solution: $2(y - 4) + 1 > 4y - 5$.

Solution: Distribute the 2 through the parentheses and simplify.

$$2y - 8 + 1 > 4y - 5$$
$$2y - 7 > 4y - 5$$

It's very important to move the variable terms to the left side of the inequality and shove everything else on the right side. In this problem, that means subtracting $4y$ from and adding 7 to both sides of the inequality. It's basically the same thing you'd do if the > sign were an =.

$$-2y > 2$$

Divide both sides by –2, remembering to reverse the inequality sign:

$$y < -1$$

Graphing this inequality is pretty easy. Start with a number line and mark the constant with either a hollow or a solid dot. (The inequality signs ≤ and ≥ translate into solid dots, since they tell you the numbers next to them should be included as part of the graph. The > and < signs, on the other hand, require open dots.) Finally, draw a dark arrow that points in the same direction as the inequality sign in your solution.

In this problem, the inequality sign is <, so place an open dot at –1, and draw a dark arrow that points left, as in Figure 2.2.

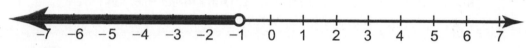

Figure 2.2

Any number less than –1, when substituted for x, *will make the inequality* 2(y – 4) + 1 > 4y – 5 *true.*

You've Got Problems

Problem 5: Solve the inequality and graph the solution: 5y ≤ 9(y + 2).

Absolute Value Inequalities

Absolute value inequalities are the divas of the mathematical world. Once they were low maintenance and the two of you got along effortlessly ("Hey, these act almost like equations. I can remember to reverse the inequality sign every once in a while, no problem!"). Seemingly overnight, however, they changed dramatically. Now they'll drink only certain kinds of water at very specific temperatures and refuse to eat any except the yellow M&Ms. (Guess who gets to open up the bag and sort them out? That's right: you.) When it comes to math, these divas require two completely different solution techniques, one for each possible way the inequality sign could point.

Less Than Inequalities

If an absolute value inequality contains either a < or a ≤ sign, you'll have to create a *compound inequality* to solve it. That means instead of one constant and one inequality sign, now you'll have two of each. Here are the specific steps you need to follow:

Talk the Talk

A **compound inequality** has two boundaries surrounding a variable expression: $a < x < b$. This can be read "a is less than x, which is less than b" or "x is between a and b." You can also write $a < x < b$ in interval notation: (a,b).

1. **Isolate the absolute value expression on the left side of the inequality.** You should end up with something like this: $|ax + b| < c$, where a, b, and c are real numbers.

2. **Rewrite as a compound inequality, removing the absolute value signs.** Turn the inequality $|ax + b| < c$ into $-c < ax + b < c$. Here's how: Drop the absolute value bars, add a matching inequality sign on the left side, and then to the left of that write the opposite of the original constant.

3. **Solve the compound inequality for x.** Don't forget that you have to add, subtract, multiply, or divide from *all three* parts of the compound inequality.

In case you're a bit fuzzy on these steps, here's an example to help you nail it down.

Example 6: Solve the inequality and graph its solution: $|3x - 2| + 5 \le 13$.

Critical Point

Some people prefer to graph inequalities using parentheses and brackets on the number line instead of open and closed dots because it makes the graphs look more like interval notation. For instance, Figure 2.3 could have a [at −2 and a] at $\frac{10}{3}$ instead of the dots.

Solution: Subtract 5 from both sides of the inequality to isolate the absolute values: $|3x - 2| \le 8$. Now rewrite this as a compound inequality by tacking on the opposite of 8 (which, of course, is −8) and another $\le$ symbol to the left of the inequality: $-8 \le 3x - 2 \le 8$. (Don't forget to drop the absolute value bars at this point as well.)

To solve for x, you must add 2 to all three parts of the inequality and then divide them all by 3:

$$-6 \le 3x \le 10$$
$$-2 \le x \le \frac{10}{3}$$

The graph of a compound inequality consists of the two dots marking the endpoints (remember to use solids dots only if the inequality contains "or equal to"; otherwise, use hollow dots) and a thick line connecting those dots, as in Figure 2.3.

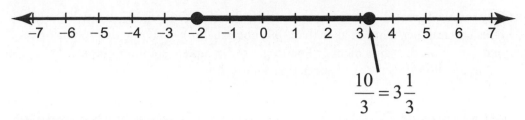

Figure 2.3

The graph of possible solutions for Example 6: Any number on the closed interval $\left[-2, \frac{10}{3}\right]$ will make the inequality $|3x - 2| + 5 \leq 13$ true.

Greater Than Inequalities

The solution for an absolute value inequality containing either > or ≥ consists of two separate, nonoverlapping intervals. Any number in either of those intervals is a possible solution. Here's how to knock these diva problems down a peg:

1. **Isolate the absolute value expression on the left side of the inequality.** This is the same initial rule you had to follow for "less than" inequalities.

2. **Write two separate *non–absolute value* inequality statements, based on the original one.** The first one's easy: Just drop the absolute value bars. To get the other one, reverse the inequality sign and take the opposite of the constant. (To be mathematically accurate, you should always write *or* between the two inequalities.) So, $|ax + b| \geq c$ would be rewritten as $ax + b \geq c$ or $ax + b \leq -c$.

3. **Solve the inequalities separately, but graph them both on the same number line.** Even though the combined graph looks like it has two arms, it is correct.

Just like its "less than" counterpart, "greater than" absolute value inequalities require their expressions to be completely rewritten before you can even begin to solve them.

Example 7: Solve the inequality and graph its solution: $2|x + 1| > 6$.

Solution: To isolate the absolute value bars, you need to divide both sides by 2; you'll end up with $|x + 1| > 3$. Rewrite this as two non–absolute value inequalities and solve them:

$$x + 1 > 3 \qquad \text{or} \qquad x + 1 < -3$$
$$x > 2 \qquad \text{or} \qquad x < -4$$

Any number greater than (but not equal to) 2 makes the original inequality true; so does any number that's smaller than (but not equal to) –4. Draw the graphs of $x > 2$ and $x < -4$ on the same number line to get the complete solution graph for the original inequality $2|x + 1| > 6$, illustrated in Figure 2.4.

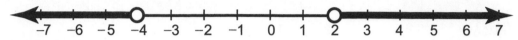

Figure 2.4

The graph of possible solutions for the inequality $2|x + 1| > 6$.

You've Got Problems

Problem 6: Solve and graph the inequalities.
    a. $|x - 5| + 1 \geq 3$
    b. $|3x| - 4 < 0$

The Least You Need to Know

◆ You can write the equation of any line, given only its slope and one of its points.

◆ Lines can be written in point-slope, slope-intercept, or standard forms.

◆ Inequalities without absolute values are easily solved, but if absolute values are present, you must first rewrite the inequality as two simpler inequalities before solving.

Practicing with Polynomials

In This Chapter

- Describing and classifying polynomials
- Basic polynomial operations
- Synthetic and long division
- Working with complex numbers

Math people get really hung up on definitions, arguably even more so than people who analyze literature for a living. I've never understood why math teachers give such careful, concise definitions of the things they teach, when almost no one understands what they are talking about.

Take, for instance, the concept of a *polynomial*. What the heck is it? Maybe this definition will help you understand: A polynomial is an expression of the form $a_0x^n + a_1x^{n-1} + a_2x^{n-2} + \ldots + a_{n-1}x + a_n$, where n is a finite integer and a_k (for any integer k such that $0 \leq k \leq n$) is a real number. Not only is that the least helpful definition in the history of time, but it probably makes you more confused!

For now, think of a polynomial as one or more clumps of things (called *terms*) added to or subtracted from one another. Those clumps are made up of numbers (called coefficients) and variables (which are often raised to whole number exponents) multiplied together.

That's all you really need to know. There will be plenty of time later to learn the rigorous definition if you decide to become a math geek like me (with all its dizzying highs, bleak lows, and more pencils than you can shake a stick at).

Polynomial Classifications

In Chapter 1, I reviewed some basic number vocabulary with you so that you'd know what an integer was and what made a number rational or irrational. Have you noticed how much time that saved us already? Now when I say *integer*, you know that I mean—a number like 7 or –5, but not 3.12598. Once we speak the same language, it saves us a whole lot of time. Can you imagine me writing "*x* has to be another one of those numbers that can be expressed as a fraction or a repeating or terminating decimal" instead of "*x* is also rational"? I'm too busy to redefine those terms every five minutes—I've got things to do; I'm sure there's a *Star Trek* rerun to watch on television somewhere!

Polynomials can be categorized just as easily as numbers, and it's just as handy a timesaver. Technically, there are two different ways to classify polynomials. The first method is based on how many terms it possesses—just count how many nonzero clumps a polynomial has, and describe it using the appropriate term from Table 3.1.

Table 3.1   Polynomial Classifications Based on the Number of Terms

Number of Terms	Classification	Example
1	Monomial	$2y^4$
2	Binomial	$9x - 1$
3	Trinomial	$x^7 + 12x + 5$

Critical Point

If a polynomial has more than one variable, find the degree by adding together the powers in each term and taking the largest sum. For instance, the degree of $2x^3y^4 - 9x^2y^6$ is 8 because $2 + 6$ (the sum of the exponents in the second term) is 8, which exceeds the first term's exponential sum of 7.

Unfortunately, this technique is only marginally useful because once a polynomial has four or more terms in it, there aren't any commonly used nicknames for it; instead you just use the generic description *polynomial*. That's why a second classification scheme is necessary. It is based not on the number of terms, but the largest exponent (called the *degree*) of the polynomial, as illustrated in Table 3.2.

Table 3.2   Classifying a Polynomial Based on Its Degree

Degree	Classification	Example
0	Constant	$5x^0$ or 5
1	Linear	$2x^1 - 1$ or $2x - 1$
2	Quadratic	$4x^2 + x - 25$
3	Cubic	$x^3 - 8$
4	Quartic	$3x^4 - x - 1$
5	Quintic	$3x^5 + 2x^4 - 7x^3 - 2$

Once again, there's an upper limit here. Other categories describe polynomials with powers larger than 5, but they're not very commonly used. Even so, with this and the classification scheme from Table 3.1, you'll be able to describe the vast majority of the polynomials you'll run across.

You've Got Problems

Problem 1: Classify the polynomials as precisely as possible.

   a. $8x^3$

   b. $3x^2 - 2x + 5$

   c. $7x - 3$

Adding and Subtracting Polynomials

If I gave you two buckets of fried chicken, each with a different assortment of pieces inside, and asked you to add up what you had, what would you do? (Let's assume, for the sake of argument, that you aren't a vegetarian, don't mind a gift of chicken buckets, and truly enjoy such eternal questions as "Hey, how much chicken you got there?" even though no one will probably ever saunter up to you and ask you that question.)

To reach an accurate chicken piece total, you'd have to break open those greasy treasure chests and sort the chicken, piece by piece. When you're through the first bucket, perhaps you'll find you've got four breasts, three wings, and five drumsticks (in the buckets, I mean—otherwise, call a doctor immediately). If the second bucket contains two breasts, seven wings, three drumsticks, and a random fried chicken beak, that makes for a grand total of:

(4 + 2) breasts + (3 + 7) wings + (5 + 3) drumsticks + 1 beak

= 6 breasts + 10 wings + 8 drumsticks + 1 beak

Your natural instinct is to sort the buckets into like pieces and add those pieces together. It wouldn't make much sense to add the three wings from the first group to the three drumsticks from the second and announce that you had a total of 6 "wingsticks," would it? No, you add things together only if they're alike.

Talk the Talk

Two polynomial terms with matching variables are called **like terms** and can be added to (or subtracted from) one another. For instance, $5x^3y^2$ and $-9x^3y^2$ are like terms because their variables, x^3y^2, match exactly. Therefore, $5x^3y^2 + (-9x^3y^2) = -4x^3y^2$.

The same thing holds true for adding and subtracting polynomials. (It's time to bring this back to math because there's only so much you can write about chicken sums before you have to commit to an entire book about chicken, and I don't think I'm quite ready for that.) Just like you sorted the chicken pieces anatomically and grouped by like parts, you sort the terms of a polynomial and group them according to the variables you find. Here's the key: You can add two variable terms together only if they have *like terms*, which means they have variables that match exactly, including their degrees.

Example 1: Simplify the expression by combining like terms:

$$(x^3 + 6x^2 - 3x + 4) - 2x(x^2 + 8x - 5).$$

Solution: Start by distributing $-2x$ through the second set of parentheses. Remember to add exponents when you multiply variable expressions with the same base (so $-2x \cdot x^2 = -2 \cdot x^{1+2} = -2x^3$).

$$x^3 + 6x^2 - 3x + 4 - 2x^3 - 16x^2 + 10x$$

According to the commutative property of addition, you can move the like terms so they're next to one another (just like you'd move matching chicken pieces into piles).

$$x^3 - 2x^3 + 6x^2 - 16x^2 - 3x + 10x + 4$$

Now combine the like terms.

$$x^3 - 2x^3 = -x^3 \qquad 6x^2 - 16x^2 = -10x^2 \qquad -3x + 10x = 7x \qquad 4 + 0 = 4$$

The final answer is the sum of all these like term results:

$$-x^3 - 10x^2 + 7x + 4$$

> **You've Got Problems**
>
> Problem 2: Simplify the expression: $-3x(4x - 3y + 1) + 5y(-2x + y + 2)$

Multiplying Polynomials

The most common algebraic error of all time is a classic mistake of incorrect binomial multiplication. Let's say you were asked to multiply $(a + b)(x + y)$. What's the answer? If you said $ax + by$, then you're yet another victim of this tragic arithmetic misconception. Don't worry—there's bound to be a support group in your area to help you work through the confusion and pain; I recommend the nonprofit organization *MisUnderstanding Polynomial Products Every Time (M.U.P.P.E.T.)*.

To help fight this epidemic, most algebra teachers introduce the FOIL technique, a trick to help you remember how to multiply two binomials properly. The name FOIL is an acronym for four words: First, Outside, Inside, and Last, all pairs of things from the problem that you need to multiply together. In the product $(a + b)(x + y)$, here are the pairs FOIL refers to:

+ **First.** The first term in each binomial (a and x)

+ **Outside.** The terms at the far left and right edges of the problem (a on the left and y on the right)

+ **Inside.** The two terms nestled in the middle of the problem (b and x)

+ **Last.** The last term in each binomial (b and y)

To calculate the product $(a + b)(x + y)$, multiply each pair of numbers together that I just described and add the result:

$$ax + ay + bx + by$$

That's a whole lot different than that wrong answer of $ax + by$, isn't it? Sure that was part of the answer, but there are two whole extra terms in there. Behold, the admirable work of the FOIL technique is done—there's one less M.U.P.P.E.T. in the world.

There's only one problem: The FOIL technique doesn't work unless you're multiplying binomials. While that wasn't too big a deal in algebra (most or all of the polynomial multiplication you did back then involved only binomials), you need to know how to multiply other polynomials as well. You'll use a method I call "extended distribution"; the good news is that it's not very hard, and you can even use it to multiply *any* two polynomials together.

Extended multiplication boils down to this rule of thumb: Multiply each term in the left polynomial (one at a time) by each term in the right polynomial. Repeat this for each term in the left polynomial, and then add all the products you get.

Example 2: Multiply the polynomials and express your answer in simplified form.

a. $(2x - y)(x + 5y)$

Multiply the first term in the left binomial ($2x$) by each term in the right binomial.

$$2x(x) + 2x(5y) = 2x^2 + 10xy$$

Now do the same thing with the remaining term in the left binomial.

$$(-y)(x) + (-y)(5y) = -xy - 5y^2$$

Add the results together and simplify.

$$2x^2 + 10xy - xy - 5y^2$$
$$= 2x^2 + 9xy - 5y^2$$

This is the same result you get with the FOIL method.

b. $(x - 3)(x^2 - 2x + 1)$

Distribute first the x and then the -3 from the left polynomial through each term in the right polynomial.

$$x(x^2) + x(-2x) + x(1) + (-3)(x^2) + (-3)(-2x) + (-3)(1)$$
$$= x^3 - 2x^2 + x - 3x^2 + 6x - 3$$

Don't forget to combine like terms.

$$x^3 - 5x^2 + 7x - 3$$

You've Got Problems

Problem 3: Calculate the product and simplify: $(-2a + 5b)(a^2 + 7ab - 4b^2)$.

Long Division of Polynomials

When I was first introduced to long division in the fourth grade, I was a bit intimidated. This new technique was so foreign that it even required its own symbol, a big horizontal line with a parenthesis glued to the left side of it. It took me a while to figure out not only what I was supposed to do, but where I was supposed to write things.

Polynomial long division works almost the same way as the long division of integers, so what I'm about to review with you will probably be vaguely familiar. However, of all the polynomial operations, students usually find division the hardest, so rather than just list the steps to follow and then do an example, I am going to do an example first and list the steps as I go along. That way, I can explain them in the context of a problem so that they'll make more sense.

Example 3: Calculate the quotient: $(2x^3 - x^2 + 5) \div (x^2 + 3x - 1)$.

Solution: The first thing to do is rewrite the problem so that the _dividend_ is inside the long division symbol and the _divisor_ is out front: $x^2 + 3x - 1 \overline{)2x^3 - x^2 + 5}$. Be sure to write both in descending powers of x.

You might have noticed that the dividend is missing an x term. It has a term to the third power and one to the second power, but then it jumps right to a constant 5 (a term with x to the 0 power, if you remember your polynomial classifications from the beginning of the chapter).

> **Talk the Talk**
>
> The division problem $a \div b$ can be rewritten as $b\overline{)a}$; b is called the **divisor**, and a is called the **dividend**.

So much of long division depends on where things are placed that you _have_ to insert a placeholder for that missing x term: $0x$. You're really just adding 0, so it doesn't change the value of the dividend (remember, 0 is the additive identity). You would have had to do the same thing if the divisor had missing x terms as well.

$$x^2 + 3x - 1 \overline{)2x^3 - x^2 + 0x + 5}$$

The setup is complete. Now ask yourself this: "What do I have to multiply x^2 by to get $2x^3$?" In other words, figure out what to multiply the leftmost term in the divisor by in order to get the leftmost term in the dividend. Since $x^2 \cdot \mathbf{2x} = 2x^3$, the answer to your self-posed question is $2x$. Write this above its like term in the dividend (the suddenly handy $0x$).

$$\begin{array}{r} 2x \\ x^2 + 3x - 1 \overline{)2x^3 - x^2 + 0x + 5} \end{array}$$

Multiply that perched $2x$ by each term in the divisor. Here's the tricky part: Put the *opposite* of each result below its like term in the dividend. For instance $2x \cdot 3x = 6x^2$, so write the opposite ($-6x^2$) below its like term in the dividend ($-x^2$). Combine the like terms, and when you're done, pull the next term in the dividend (the constant 5) down to join the newly combined terms.

$$
\begin{array}{r}
2x \\
x^2 + 3x - 1{\overline{\smash{\big)}\,2x^3 - x^2 + 0x + 5}} \\
\underline{-2x^3 - 6x^2 + 2x} \\
-7x^2 + 2x + 5
\end{array}
$$

Critical Point

You write the opposites of the products beneath the division sign because, technically, you're subtracting them.

The question to ask yourself now is "What do I multiply x^2 (still the leftmost term in the divisor) by to get $-7x^2$ (the leftmost term of the bottom line in the problem)?" The answer is -7, and it (like the last such answer) is written above its like term in the division symbol. Repeat the process of multiplying it by each term in the divisor, writing down the opposites below, and then combining the nicely stacked like terms.

$$
\begin{array}{r}
2x - 7 \\
x^2 + 3x - 1{\overline{\smash{\big)}\,2x^3 - x^2 + 0x + 5}} \\
\underline{-2x^3 - 6x^2 + 2x} \\
-7x^2 + 2x + 5 \\
\underline{7x^2 + 21x - 7} \\
23x - 2
\end{array}
$$

Since there are no more terms in the dividend to "drop down" (as you did with the 5 earlier), you know you're finished dividing. The expression above the division symbol is the quotient, and the bottommost line is the remainder. Your final answer is the quotient plus the fraction whose numerator is the remainder and whose denominator is the divisor:

$$
2x - 7 + \frac{23x - 2}{x^2 + 3x - 1}
$$

You can always check your work by multiplying the quotient by the divisor and then adding the remainder; you should get the original dividend.

$$(\text{quotient})(\text{divisor}) + (\text{remainder}) = (\text{dividend})$$
$$(2x-7)(x^2+3x-1)+(23x-2) = 2x^2-x^2+5$$
$$2x^3-x^2-23x+7+(23x-2) = 2x^2-x^2+5$$
$$2x^3-x^2+5 = 2x^2-x^2+5$$

You've Got Problems

Problem 4: Calculate the quotient: $(x^3 - 9x - 4) \div (x - 3)$.

Synthetic Division

Now that (I hope) you understand long division, I can be fully honest with you—I really dislike it. There's a lot to remember: switching the signs at the correct time, when to write the numbers above the division symbol and when they go below, ensuring that you "drop down" another dividend term each time. Any little mistake in any of these steps means you get the whole problem wrong.

Well, there's no use complaining unless there's a better way, right? The good news is that there is a better way: *synthetic division,* a shortcut method you can use to divide any kind of polynomial by a linear binomial. Synthetic division calculates the quotient of a polynomial dividend and a binomial divisor of the form $x + a$, using only the co-efficients and constants of the expressions. Since no variables are needed, the process is easier and quicker. (The divisor has to be in the form $x + a$, where a is a positive or negative rational number.)

CAUTION

Kelley's Cautions

Synthetic division works only if the divisor is a linear binomial that looks like $x + a$ (although a could be negative). If you have a divisor like $4x + 3$, divide both terms by x's coefficient to get it into the form you need for synthetic division. In other words, $\frac{4x}{4} + \frac{3}{4} = x + \frac{3}{4}$. This tells you to rewrite the divisor $4x + 3$ as $x + \frac{3}{4}$.

Example 4: Calculate $(2x^3 - x + 3) \div (x + 4)$.

Solution: Write the *opposite* of the divisor's constant in a small box. (In this problem, the divisor's constant is 4, so you write –4 in the box.) Then list all the dividend's coefficients in order, from the highest power of x to the lowest. If there are any missing x exponents, make sure to insert a 0 placeholder. In this case, $2x^3 - x + 3$ has no x^2 term, so think of it as $2x^3 + 0x^2 - x + 3$ when you list those coefficients.

Leave some space below that row of numbers and then draw a horizontal line. Below that line, rewrite the leftmost coefficient (that's not in the box)—the number 1, in this problem. You should end up with the following result:

$$\begin{array}{r|rrrr} -4 & 2 & 0 & -1 & 3 \\ \hline & 2 & & & \end{array}$$

Multiply the box number by the 2 below the line ($-4 \cdot 2 = -8$), and write the result below the 0. Then add that column ($0 + (-8) = -8$) and write the result underneath, below the line.

$$\begin{array}{r|rrrr} -4 & 2 & 0 & -1 & 3 \\ & & -8 & & \\ \hline & 2 & -8 & & \end{array}$$

Keep repeating this process: Multiply the box number by the rightmost number below the line, write the result above the line in the column immediately to the right, and then sum the numbers in the column, writing the result below the line. Repeat until there are no more blank spots below the original row of coefficients.

$$\begin{array}{r|rrrr} -4 & 2 & 0 & -1 & 3 \\ & & -8 & 32 & -124 \\ \hline & 2 & -8 & 31 & -121 \end{array}$$

All the numbers below the line (except the final one on the right) are the coefficients of the quotient; the last number is the remainder. Your answer, of course, needs to have variables in it, and the degree of your quotient will be exactly one less than the degree of your original dividend. Since $2x^3 - x + 3$ has degree 3 (since its highest exponent is 3), the quotient must begin with x to a power of 2.

$$2x^2 - 8x + 31 + \frac{-121}{x+4}$$

Once again, the remainder is added on at the end, as the numerator of a fraction whose denominator is the original divisor. In this problem (since the remainder is negative), it is also correct to write $2x^2 - 8x + 31 - \dfrac{121}{x + 4}$; you're allowed to move the negative sign in front of the remainder. If you want, check your answer the same way you did with long division:

$$(\text{quotient})(\text{divisor}) + (\text{remainder}) = (\text{dividend})$$

You've Got Problems

Problem 5: Use synthetic division to calculate $(x^4 - 6x^3 + 2x^2 - 5) \div (x - 2)$.

Complex Numbers

In Chapter 1, I reintroduced you to the amazing little number $i = \sqrt{-1}$ (referring to complex numbers). Just like an ant (which can supposedly lift hundreds of times its body weight without breaking an ant sweat), i is just a little teeny letter, but it helps you to do powerful things, such as simplify negative radicals.

Let's say you're given the radical expression $\sqrt{-16x}$. To simplify this, first rewrite it as $i\sqrt{16x}$. (Erase that troublesome negative sign inside the radical and pop an i out in front.) Now that the radical is positive, you should have no trouble simplifying to end up with $i \cdot 4\sqrt{x}$, or $4i\sqrt{x}$. Because the final answer has an i, it's called an *imaginary number*.

If you think back to Chapter 1, a complex number is made by adding a real number to an imaginary number: $a + bi$. You probably won't be asked to do more than add, subtract, multiply, and divide complex numbers in precalculus, so let's make sure you know how to do exactly that. For the most part, they act like binomials (which, in case you're curious, is why I stuck them in this chapter, even though they are technically not binomials).

Talk the Talk

You can identify an **imaginary number** (such as $-5i$ or $2i\sqrt{3}$) because it contains i, which is understood to have a value of $\sqrt{-1}$. Imaginary numbers are an essential part of complex numbers, which contain both a real number part and an imaginary number part.

Example 5: Simplify the complex expressions:

a. $3(2 - 4i) - 5(1 + 7i)$

Don't stress out that the variable in the expression is i instead of x; you'll distribute the exact same way:

$$3(2) + 3(-4i) + (-5)(1) + (-5)(7i)$$
$$= 6 - 12i - 5 - 35i$$

The 6 and −5 are like terms, as are −12i and −35i, so combine them to get $1 - 47i$.

b. $(4 - 2i)(-5 + 3i)$

Use the FOIL or extended distribution method to multiply.

$$-20 + 12i + 10i - 6i^2$$

If $i = \sqrt{-1}$, then $i^2 = \left(\sqrt{-1}\right)^2$, so $i^2 = -1$. So, substitute −1 for i^2 and simplify:

$$-20 + 12i + 10i - 6(-1) = -14 + 22i$$

c. $(4 + 9i) \div (-3 - 6i)$

Rewrite this as a fraction (the first complex number is the numerator and the second is the denominator), and multiply both the top and bottom of the fraction by the *conjugate* of the denominator, which is −3 + 6i.

$$\frac{4+9i}{-3-6i} \cdot \frac{-3+6i}{-3+6i} = \frac{-12+24i-27i+54i^2}{9-18i+18i-36i^2} = \frac{-66-3i}{45}$$

Talk the Talk

The **conjugate** of $a + bi$ is $a - bi$, and vice versa; just change the sign between the terms.

You can simplify further by dividing everything by the common factor of 3 to get $\frac{-22-i}{15}$. Your teacher may want you to write this in official $a + bi$ complex number form, which would be $-\frac{22}{15} - \frac{1}{15}i$.

You've Got Problems

Problem 6: If $c = 1 - 6i$ and $d = 4 + 3i$, calculate $c + d$, $c - d$, $c \cdot d$, and $c \div d$.

The Least You Need to Know

- Polynomials are made up of clumps, called terms, that contain coefficients and variables raised to powers.

- You can add or subtract only like terms in a polynomial, which contain the exact same variables.

- Whereas long division allows you to calculate the quotient of any two polynomials, synthetic division requires that the divisor be a linear binomial of the form $x + a$ or $x - a$.

- Complex numbers are a lot like binomials because they are governed by similar rules and restrictions.

Chapter 4

Factoring Polynomials

In This Chapter

- ◆ Recognizing greatest common factors
- ◆ Factoring out monomials and binomials
- ◆ Common factoring patterns to memorize
- ◆ Turning trinomials into products of binomials

Now that you're really good at multiplying polynomials, let's see how good you are at "unmultiplying" them. That's what the process of *factoring* is all about. In Chapter 3, you learned how to turn $(x - 3)(2x + 1)$ into $2x^2 - 5x - 3$, but in this chapter, you'll have to work in reverse, changing $2x^2 - 5x - 3$ back into $(x - 3)(2x + 1)$. Sure, that might seem easy now, but it gets a bit more tricky when I don't give you the answer ahead of time.

Of course, the easiest way to factor a polynomial would be to not multiply its component parts together in the first place. Unfortunately, since time machines have yet to be invented (and you lack the power to fly around the world so fast that it actually spins Earth backward and reverses time, saving Lois Lane from an untimely death in *Superman II*), you might as well learn to factor—either that or go ahead and invent a time machine of your own. (You'll do anything to avoid reviewing algebra, won't you?)

Greatest Common Factors

Do you remember how to find the greatest common factor of two integers? You probably learned it way back in your first algebra class, and have either forgotten how to do it or have internalized it and are able to give answers without really knowing how. Either way, it's worthwhile to stop for a minute and review how to find that mysterious greatest common factor (GCF).

Finding Greatest Common Factors of Integers

First of all, you need to know what a *factor* is. Basically, it's a quantity that divides evenly into something else. (By "dividing evenly," I mean that the answer will be an integer.) For example, 5 is a factor of 10, since 5 divides evenly into 10 (the quotient $10 \div 5$ is equal to 2, which is a nice whole number, not a fraction). On the other hand, 4 is not a factor of 10 because $10 \div 4 = \frac{5}{2}$, which is a rational number, but not an integer. Now that you have the basic language under your belt, you're ready to find greatest common factors.

Talk the Talk

If a is a **factor** of b, then the quotient $b \div a$ will have a remainder of 0, and you can say that a "divides evenly" into b. If (as in Example 1) a and b are just integers, then a is a factor of b if $b \div a$ results in an integer. The only factors of a **prime** number are 1 and the number itself.

Example 1: What is the greatest common factor of 32 and 80?

Solution: Your goal is to break each number into its *prime* factors. It's actually pretty easy to do: Just start with 32 and come up with any two numbers that have that product. I'll choose 8 · 4. (You could also have chosen 16 · 2, but you'll get the same final answer.)

32

8 · 4

Critical Point

The arrow charts in Example 1 are called factor trees because each number sprouts two branches containing its factors. If either of those numbers is not prime, it sprouts its own factor branches as well. To get the prime factorization of the top number, just multiply together all the numbers in the tree that don't have branches below them.

Since 8 and 4 are not prime numbers, rewrite them as products of two factors, just like you did for 32 a moment ago ($8 = 2 \cdot 4$ and $4 = 2 \cdot 2$).

$$
\begin{array}{c}
32 \\
\swarrow \searrow \\
8 \quad \cdot \quad 4 \\
\swarrow \searrow \quad \swarrow \searrow \\
4 \cdot 2 \quad 2 \cdot 2
\end{array}
$$

The only number left in the bottom row ($4 \cdot 2 \cdot 2 \cdot 2$) that's not prime is 4, so break it down into factors as well. To draw extra attention to them, I will box in all the prime numbers (so they have no arrows extending from their bottoms, which is a good thing, as you'd know if you've ever sat on a quiver).

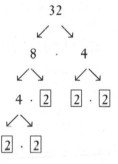

Multiply all those boxed numbers together: $2 \cdot 2 \cdot 2 \cdot 2 \cdot 2 = 2^5$; this is the fully factored (prime factorization) version of 32. Compare this to the prime factorization of 80.

$$
\begin{array}{c}
80 \\
\swarrow \searrow \\
8 \quad \cdot \quad 10 \\
\swarrow \searrow \quad \swarrow \searrow \\
4 \cdot \boxed{2} \quad \boxed{5} \cdot \boxed{2} \\
\swarrow \searrow \\
\boxed{2} \cdot \boxed{2}
\end{array}
$$

Therefore, the prime factorization of 80 is $2 \cdot 2 \cdot 2 \cdot 5 \cdot 2 = 2^4 \cdot 5$.

To calculate the greatest common factor of 32 and 80, grab the lowest exponent on every factor they have in common. In this case, 32 and 80 have only one factor in common: 2 (because 32 contains 2^5 and 80 contains 2^4). The greatest common factor is 2 raised to the lower of the powers: $2^4 = 16$. If there had been more common factors, you would have taken the smaller exponential power of each one (try the upcoming Problem 1 for an example like this).

What does it all mean? Simply put, because 16 is the greatest common factor of 32 and 80, it is guaranteed to be the biggest integer that divides evenly into both numbers.

You've Got Problems

Problem 1: Find the greatest common factor of 72 and 180.

Finding Greatest Common Factors of Polynomials

Once you can find the greatest common factors of integers, it's simple to take the next step and find the greatest common factors of terms in a polynomial. In fact, the only difference is that now you'll have to deal with some variables thrown into the mix, but the basic rule of thumb remains the same: Build the greatest common factor by keeping the lowest exponential version of each shared factor.

Example 2: Factor the polynomial $15x^3y^2 - 45xy^4z$.

Solution: Start by calculating the greatest common factor of the coefficients—they're just integers, so you can use the factor trees I showed you in the last section. You'll find that $3 \cdot 5 = 15$ is the largest integer that divides into both evenly. Now it's time to move on to the variables.

Critical Point

If every term in the polynomial you're factoring is negative, you should factor a −1 out of each as part of the greatest common factor.

Of the three variables in the polynomial (x, y, and z), only x and y are found in both terms, so z is not a candidate for the greatest common factor. Take the lowest power of x you see in the remaining terms (x instead of x^3), and do the same thing for y (use y^2 instead of y^4). Slam all the pieces together (the number and the correct exponents of x and y) to get the greatest common factor of $15xy^2$.

Once you know what the greatest common factor is, you should factor it out of each term, writing the polynomial in factored form. Write the greatest common factor, and next to it (in a set of parentheses), rewrite the polynomial with the greatest common factor divided out of it.

$$15xy^2\left(\frac{15x^3y^2}{15xy^2}-\frac{45xy^4z}{15xy^2}\right)$$

How do you simplify those fractions? The number parts are easy—just simplify those like you would any normal fraction from your prealgebra past, so $\frac{45}{15}=\frac{\cancel{15}\cdot3}{\cancel{15}}=\frac{3}{1}=3$. To simplify the variable parts, apply the formula $\frac{x^a}{x^b}=x^{a-b}$ from the section "Exponential Rules" in Chapter 1.

$$15xy^2\left(\frac{15}{15}x^{3-1}y^{2-2}-\frac{45}{15}x^{1-1}y^{4-2}z\right)$$
$$=15xy^2\left(x^2-3y^2z\right)$$

Since you've sucked out all the repetitive terms by writing the greatest common factor only once, the end result is a much more compact version of the original polynomial. It's very simple to check your work: Just distribute the $15xy^2$ back through the parentheses once again, and you should get the original expression.

$$15xy^2\left(x^2-3y^2z\right)$$
$$=15x\cdot x^2\cdot y^2-15\cdot3\cdot x\cdot y^2\cdot y^2\cdot z$$
$$=15x^3y^2-45xy^4z$$

CAUTION **Kelley's Cautions**

Remember, your answer for Example 2 is the polynomial in factored form. You can distribute to check your answer, but don't turn in the redistributed expression as your final answer. Otherwise, your answer would be an exact copy of the original problem, and that wouldn't make much sense!

You've Got Problems

Problem 2: Factor the polynomial $-18a^3b^5c^2-9a^2b^3c^7-6a^3bc^5$.

Factoring by Grouping

When I was a student, I never heard of "factoring by grouping"; it wasn't a term that my teachers used a lot, and the phrase had absolutely no meaning for me. Even throughout my college years as a math major (and remarkably unfashionable dresser—how many pairs of red sweatpants does one person need in order to prove that he has no fashion sense of his own?), the phrase "factoring by grouping" never really came up. We math geeks never discussed them at our lame math parties, amid rollicking games of "Spin the Protractor" and "Truth, Dare, or Quadratic Equation."

Critical Point

If you're asked to factor a polynomial containing four terms (no two of which are like terms), it's likely that you'll need to factor by grouping.

It wasn't until I started teaching that I learned what this term actually meant, and, lo and behold, I had been doing it all along. I just never knew it had its own fancy name. It's like finding out that your friend Barry's real name is Bartholomew Alistair Ramirez Michelangelo Baryshnikov III. Pretty fancy, but to you, he'll always be just Barry, the guy who threw up in the punch bowl at your sister's wedding.

Factoring by grouping is a fancy way of saying "A greatest common factor is allowed to contain binomials as well as simple monomial bits such as numbers and letters." For instance, in the expression $x^2(y-1)^3 + x(y-1)^2$, there are two common factors: x and $(y-1)$. It doesn't matter that one of them is nothing but a letter (x) and the other is the big chunky binomial ($y-1$)—they're separate factors, so you should compare their powers and take the lesser one to construct the greatest common factor, just like you've done all along: $x(y-1)^2$.

Example 3: Factor the polynomial $3x^3 - 12x^2 + 2x - 8$.

Solution: As I mentioned in the nearby "Critical Point," you should lean toward factoring by grouping to solve this problem because it contains four terms with no greatest common factor. Start by rewriting the expression by splitting it into two pieces, pairing the first two terms and the last two terms together. (This forced grouping of pairs is where the technique gets its name.)

$$(3x^3 - 12x^2) + (2x - 8)$$

Notice that each pair has its own greatest common factor, so rewrite those binomials in factored form:

$$3x^2 (x - 4) + 2(x - 4)$$

This is the key step to factoring by grouping: The two clumps now feature a matching binomial that should be factored out. In this problem, you should factor $(x - 4)$ out of each term, and you're finished:

$$(x - 4)(3x^2 + 2)$$

Kelley's Cautions

If you're factoring by grouping and find that you don't end up with matching binomials, but instead you have binomials that are opposites of one another, factor −1 out of the second binomial and you'll be good to go. For example, if you have $2y(y - 4) + 7(-y + 4)$, it becomes $2y(y - 4) - 7(y - 4)$.

You've Got Problems

Problem 3: Factor the polynomial: $10xw - 15x - 12wy + 18y$.

Common Factor Patterns

I'm not against good old-fashioned work, but I love shortcuts as much as anyone. When it comes to factoring, you're not always required to do a ton of work. Sometimes you can just look at the problem and immediately know the answer. There are three very common patterns for factoring, and if you memorize these patterns, it will dramatically decrease the amount of time and frustration you'll expend with polynomials.

Here are the three most common factoring patterns and what they mean:

Kelley's Cautions

Remember that there is no formula to factor the sum of perfect squares. Many students assume that $a^2 + b^2 = (a + b)(a + b)$, but that is not true! In fact, this is the M.U.P.P.E.T. misconception from the section "Multiplying Polynomials" in Chapter 3, just written in reverse—and it's still wrong.

◆ **Difference of perfect squares, $a^2 - b^2$ = $(a + b)(a - b)$.** Any two perfect squares that are subtracted can be instantly factored into two binomials, one of which adds their square roots and one of which subtracts them. For example, $x^2 - 25 = (x + 5)(x - 5)$ and $16y^2 - 9 = (4y + 3)(4y - 3)$.

♦ **Sum of perfect cubes, $a^3 + b^3 = (a + b)(a^2 - ab + b^2)$.** This formula is a bit more complex than its predecessor, but you should still memorize it. You use it, as its name suggests, to factor any two perfect cubes added together, such as $27x^3 + 8$.

It makes your work a bit easier if you rewrite the expression in terms of the actual cube roots: $27x^3 + 8 = (3x)^3 + (2)^3$. If you compare this to the formula $a^3 + b^3$, $3x$ is the a term and $b = 2$. Just plug those into the formula to factor $27x^3 + 8$ instantly:

$$a^3 + b^3 = (a + b)(a^2 - ab + b^2)$$
$$(3x)^3 + (2)^3 = (3x + 2)\left((3x)^2 - (3x)(2) + (2)^2\right)$$
$$27x^3 + 8 = (3x + 2)(9x^2 - 6x + 4)$$

♦ **Difference of perfect cubes, $a^3 - b^3 = (a - b)(a^2 + ab + b^2)$.** This is the sister formula to the sum of perfect cubes, differing by only two signs, if you look closely.

Critical Point _____

The first thing you should do in *any* factoring problem is to check for a greatest common factor; if it exists, factor it out immediately. You can then progress on to other techniques, such as factor patterns or the methods you'll learn at the end of this chapter. Otherwise, it may be impossible to do the problem at all, and even if you manage it, your answer probably won't be fully factored anyway.

When you're done, make sure to factor all of your answers completely.

I know you hate memorizing things; I do, too. However, you'll need to be incredibly smart to figure out the cube formulas on the fly. Imprinting the formulas deep within your brain can make an otherwise impossible problem almost trivially easy, which is always a good thing.

You've Got Problems

Problem 4: Factor the polynomials completely:
   a. $3x^4 - 48y^4$
   b. $r^3 - 125$

Factoring Quadratic Trinomials

The easiest quadratic polynomials to factor have a *leading coefficient* of 1 (I'll deal with the harder ones in the final section of this chapter). You'll be given a trinomial such as $x^2 + ax + b$, and your goal will be to factor it into two binomials: $(x + ?)(x + ?)$. Obviously, your first task is to figure out what the heck belongs in place of those question marks—there are two numbers unique for every polynomial $x^2 + ax + b$, and they have these characteristics:

1. **They will add up to a.** In other words, when you add those mystery numbers together, you should get the coefficient right smack in the middle of the polynomial.

2. **They will multiply to equal b.** Not only does their sum show up in the polynomial $x^2 + ax + b$, but so does their product; it must be equal to the constant.

> **Talk the Talk**
>
> The **leading coefficient** of a polynomial is the coefficient of the term with the highest exponent. For example, the leading coefficient of $4x^3 + x^2 - 3x^5 + 9$ is -3. It's "leading" the polynomial because it would be first in line if you were to write the polynomial in descending powers of x.

Once you solve the mystery of the missing values, use them to replace the question marks in $(x + ?)(x + ?)$ and instantly factor the trinomial. (Of course, if the trinomial contains a variable other than x, as in $y^2 - 3y - 4$, your final answer should contain the actual variable used in the problem.)

Example 4: Factor the polynomials completely.

a. $x^2 - 9x + 18$

Solution: What two numbers add up to –9 but multiply to give you 18? Notice that the product is positive, so the two numbers must have the same sign (two numbers with different signs always have a negative product). This leads you to another question: Are both numbers positive or negative? Well, since their sum is –9, both must be negative.

The only two negative numbers that fulfill the necessary requirements are –6 and –3: $-6 + (-3) = -9$, and $(-6)(-3) = 18$. Use them to replace the question marks in $(x + ?)(x + ?)$, and the polynomial is factored: $(x - 3)(x - 6)$. It doesn't matter which you write first because the end result is the same, according to the commutative property of multiplication.

b. $x^2 - 10x + 25$

Solution: You're looking for two numbers that add up to –10 and multiply to give you 25. Will the numbers 5 and 5 work? (You're not breaking any rules if the mystery numbers are equal). Sure, $5 \cdot 5 = 25$, but $5 + 5 \neq -10$, so those don't work. However, notice that –5 and –5 do work: $(-5)(-5) = 25$, and $-5 + (-5) = -10$. So, the factored form of the polynomial is $(x - 5)(x - 5)$ or $(x - 5)^2$.

c. $2w^2 - 8w - 90$

Solution: The leading coefficient of this polynomial is not 1, so you cannot yet use our mystery number technique. However, you can factor out the greatest common factor of 2 to get $2(w^2 - 4w - 45)$. Inside the parentheses, the leading coefficient is now 1, so you can proceed. (Don't worry about that 2 hanging out there in limbo. Now that it's factored out, you don't have to do anything else with it except keep carrying it out in front of each step of the problem.)

What two numbers multiply to –45 and add to –4? The constant is negative, so the numbers will have different signs. Because their sum is negative, you know that the bigger of the two numbers will be negative. The only two numbers that work are –9 and 5: $-9 + 5 = -4$, and $(-9)(5) = -45$. So, your final answer is $2(w - 9)(w + 5)$. Don't forget to write the 2 that you factored out earlier in front of the binomial factors.

You've Got Problems

Problem 5: Factor the polynomials completely:
  a. $y^2 + 7y + 12$
  b. $3x^2 + 9x - 84$

The Bomb Method of Factoring Trinomials

If you're trying to factor a quadratic trinomial whose leading coefficient is not 1, you can use another method to factor, a method officially called "factoring by decomposition" but one that I call the "Bomb Method." It causes an explosion in the middle term of a quadratic trinomial, cracking it into two pieces in a very precise fashion. This controlled burst actually results in a polynomial with four terms that can then be factored by grouping.

Here are the steps you'd follow to try to factor a polynomial that looks like $ax^2 + bx + c$:

1. **Find two numbers that add up to b and multiply to give you $a \cdot c$.** This is similar to the previous trinomial factoring method, but now the leading coefficient comes into play.

2. **Rewrite the b term as the sum of the two numbers you found in step 1.** This step plants the bomb in the polynomial.

Critical Point

If you're factoring $ax^2 + bx + c$ and $a = 1$, the Bomb Method still works. It actually turns into the factoring technique you learned in the previous section because $a \cdot c = c$.

3. **Factor by grouping.** Notice that there aren't any new skills involved in the Bomb Method. It just tweaks a few techniques you already know and, in so doing, makes a hard problem much easier.

Not many people know about the Bomb Method, so they don't have any real strategy for approaching tough factoring problems. They just experiment with binomials until they get the right answer. I'd rather have a plan going in, which is why I'm such a big fan of this technique. (If I were a much cooler person, I'd even call it "Da Bomb," but I'm not, so forget I said that.)

Example 5: Factor $6x^2 - x - 2$.

Solution: Your first goal is to find two numbers that have a sum of -1 (the x coefficient) and a product of -12 (since $6(-2) = 12$). The only two such numbers are -4 and 3. Write their sum in place of the x coefficient:

$$6x^2 + (-4 + 3)x - 2$$

Since $-4 + 3 = -1$, you haven't changed the polynomial at all. (Now you can see why the -1 sum was so important.) Distribute the x to continue.

$$6x^2 - 4x + 3x - 2$$

Even though the last two terms have no greatest common factor other than 1, you still need to factor by grouping. This is one of the few times you'll actually factor out a factor of 1.

$$2x(3x - 2) + 1(3x - 2)$$
$$= (3x - 2)(2x + 1)$$

You've Got Problems

Problem 6: Factor $3x^2 + 2x - 8$.

The Least You Need to Know

◆ The greatest common factor is the largest thing that divides evenly into every term of a polynomial.

◆ You can factor the difference of perfect squares, the sum of perfect cubes, and the difference of perfect cubes.

◆ A quadratic is slightly harder to factor if its leading coefficient is not 1.

Rational Expressions and Weird Inequalities

In This Chapter

- ◆ Performing basic operations on rational expressions
- ◆ Calculating a least common denominator
- ◆ Solving equations containing fractions
- ◆ Solving rational and quadratic inequalities

To succeed in math, you'll need many of the same skills required of an ambassador to the United Nations. Different nations have different customs, and a comprehensive knowledge of those cultural differences is a job prerequisite. Whereas a healthy belch after a meal may, in one country, communicate your appreciation and thanks to the chef, my loud belching after meals will communicate only gastrointestinal distress and will humiliate my wife, who should have known better than to take me someplace nice.

In the world community of mathematics, you have to be culturally sensitive as well. For instance, integers have no hang-ups about getting added to one another; they're like hippies in the era of free love. However, in the realm of rational numbers (which are fractions, as you may recall from

Chapter 1), citizens are very tense about the whole adding process. Therefore, they have very specific laws governing addition and subtraction with their members— common denominators are mandatory, and if you forget it, there's a minimum fine of $\$\frac{1250}{42}$ or up to $\frac{21}{5}$ years in prison.

In this chapter, you'll reacquaint yourself with the rules of rational expressions. You'll find that even though their numerators and denominators may contain variables, the same laws that govern plain old vanilla rational numbers hold sway with the more complicated rational expressions as well.

Adding and Subtracting Rational Expressions

As I mentioned in the preceding bizarre ambassador metaphor, rational expressions must possess *common denominators* before their numerators can be combined. Although two rational expressions can share an infinite number of equivalent denominators, your best bet is to calculate the *least common denominator* (LCD) and use it to combine the fractions.

The Least Common Denominator

Remember how you found the greatest common factor in Chapter 4? The first thing you did was a complete prime factorization using a factor tree. Next, you compared the results to see what factors each term had in common. You used the lowest exponential value of each common term to construct the greatest common factor.

Talk the Talk

If two fractions have the exact same quantity in the denominator, they are said to have **common denominators**. The smallest possible common denominator for a group of rational expressions is called the **least common denominator**. Calculating it requires a technique similar to the one you used to calculate greatest common factors in Chapter 4.

You'll need to use prime factorization again to determine the least common denominator. However, there are two major differences in how you'll use it:

1. **Use all the factors, not just the common ones.** Every single factor of every expression involved must appear somewhere in the least common denominator.

2. **Instead of the smallest exponential power of each factor, use the largest.** If a factor is shared, you should raise it to the highest power you see in the expressions, not the lowest, as you did in finding the greatest common factor.

Once you have found the least common denominator, you can rewrite all of the given rational expressions so that it appears in each one, finally allowing addition and subtraction of those expressions.

Example 1: What is the least common denominator of the rational expressions $\dfrac{7}{3x}$, $\dfrac{x-5}{27(2x+1)^4}$, and $\dfrac{x+2}{x^3(8x+4)^2}$?

Solution: Start by finding the prime factorization of each denominator:

◆ **$3x = 3x$.** The first denominator is already prime. It cannot be factored any further.

◆ **$27(2x + 1)^4 = 3^3(2x + 1)^4$.** The constant 27 can be factored using a factor tree to give you 3^3, but the terms of the binomial $(2x + 1)$ share no greatest common factor (other than 1), so the binomial is left alone.

◆ **$x^3(8x + 4)^2 = x^3 \cdot 4^2(2x + 1)^2 = x^3 \cdot 2^4(2x + 1)^2$.** The binomial has a greatest common factor of 4 that can be pulled out, but when you do so, make sure to write it as 4^2. See the nearby "Kelley's Cautions" sidebar to find out why. Finally, write the prime factorization of 4^2. Because $4 = 2^2$, $4^2 = (2^2)^2 = 2^{2 \cdot 2} = 2^4$.

CAUTION

Kelley's Cautions _____

When you pull factors out of an expression raised to a power, those factors should also be raised to that power. For instance, the binomial inside $(xy - y^2)^5$ has a greatest common factor of y, but because the binomial is raised to the fifth power, the y should be also, once factored out: $y^5(x - y)^5$.

Here are all the unique factors in the denominators (ignoring exponents for now): 3, x, $(2x + 1)$, and 2. All of these belong in the least common denominator, but you attach to each one the highest exponential value it has in all three denominators: $3^3 \cdot x^3 \cdot (2x + 1)^4 \cdot 2^4$. Multiply the constants to get the least common denominator: $432x^3(2x + 1)^4$.

You've Got Problems

Problem 1: Find the least common denominator of the rational expression
$\dfrac{x^2}{x^2 - 3x + 2} + \dfrac{x-2}{x^2 - 2x + 1} - \dfrac{5x^2}{x^2 + 3x - 10}$. (*Hint:* Factor the denominators first.)

Combining Rational Expressions

Now that you remember how to calculate a least common denominator, you need to know what the heck to do with it. Your goal is to turn every denominator in the rational expression into the least common denominator, and here's how it's done:

1. **Divide the least common denominator by the first denominator in the expression.** You can do this in your head, if you like. It doesn't have to be explicitly written in your work.

2. **If you get a remainder, multiply the numerator and denominator of that fraction by that remainder.** Since you're multiplying both parts of the fraction by the same thing, you're essentially multiplying by 1, so it doesn't change the fraction's value.

3. **Lather, rinse, and repeat.** Repeat this process for every single fraction.

4. **Write the combined numerator over the common denominator.** Now you're free to add and subtract the numerators. Just be sure to write the result over the common denominator.

I know this sounds tricky, and to be perfectly honest, there are a lot of places you can make tiny but significant mistakes, so you'll need to be careful. The good news is that, believe it or not, adding and subtracting rational expressions is one of the hardest things you'll do with them, so once you master this process, the rest of this chapter should be smooth sailing.

Example 2: Simplify the expression from the previous "You've Got Problems" (Problem 1) earlier this chapter:

$$\frac{x^2}{x^2 - 3x + 2} + \frac{x - 2}{x^2 - 2x + 1} - \frac{5x^2}{x^2 + 3x - 10}$$

Solution: Start by factoring the denominators of the rational expression.

$$\frac{x^2}{(x-2)(x-1)} + \frac{x-2}{(x-1)(x-1)} - \frac{5x^2}{(x+5)(x-2)}$$

If you've already worked out Problem 1 (or cheated and looked in Appendix A for the answer), you know that the least common denominator is $(x - 2)(x - 1)^2(x + 5)$. Divide this common denominator by each of the previous denominators, one at a time. Any pairs of quantities appearing in both the numerator and the denominator can be crossed out (because anything divided by itself is 1). Here's what you get when you divide the common denominator by the denominator of the leftmost fraction:

$$\frac{\cancel{(x-2)}\,\cancel{(x-1)}\,(x-1)(x+5)}{\cancel{(x-2)}\,\cancel{(x-1)}} = (x-1)(x+5)$$

Multiply that result, $(x-1)(x+5)$, by the top and bottom of the leftmost fraction.

$$\frac{x^2}{(x-2)(x-1)} \cdot \frac{(x-1)(x+5)}{(x-1)(x+5)} = \frac{x^2(x-1)(x+5)}{(x-2)(x-1)^2(x+5)}$$

See what happened? You altered the fraction, forcing it to switch to the least common denominator, which disguised its numerator as well. Multiply all those numerator terms together, but leave the denominator alone.

$$\frac{x^2(x-1)(x+5)}{(x-2)(x-1)^2(x+5)} = \frac{x^4 + 4x^3 - 5x^2}{(x-2)(x-1)^2(x+5)}$$

Now do the same thing for the middle and right functions, and they'll all have the same denominator.

$$\frac{x^4 + 4x^3 - 5x^2}{(x-2)(x-1)^2(x+5)} + \frac{x^3 + x^2 - 16x + 20}{(x-2)(x-1)^2(x+5)} - \frac{5x^4 - 10x^3 + 5x^2}{(x-2)(x-1)^2(x+5)}$$

CAUTION

Kelley's Cautions

If one of the fractions in a rational expression is negative, be sure to distribute that negative through each term in the numerator when you combine them. For instance, when you simplify $\dfrac{w}{4} + \dfrac{2x-3}{4} - \dfrac{y+5z}{4}$, you'll get $\dfrac{w + 2x - 3 - y - 5z}{4}$. Students often forget that $5z$ needs to be negative like the y that precedes it because the subtraction sign in front of $\dfrac{y+5z}{4}$ in the original problem affects both numerator terms.

Now that everything has the same denominator, you can combine all the numerators and simplify:

$$\frac{x^4 + 4x^3 - 5x^2 + (x^3 + x^2 - 16x + 20) - (5x^4 - 10x^3 + 5x^2)}{(x-2)(x-1)^2(x+5)} = \frac{-4x^4 + 15x^3 - 9x^2 - 16x + 20}{(x-2)(x-1)^2(x+5)}$$

If you knew how, you would factor the numerator at this point and see if the fraction could be simplified (by crossing out matching factors, like you did earlier in the problem). However, factoring a quartic polynomial is no trivial matter—I won't discuss it until Chapter 7—so you can just leave your answer like this (unless you really want to multiply out the denominator so that both parts of the fraction are fully expanded). Either way, you've done as much simplifying as you can right now.

You've Got Problems

Problem 2: Simplify the expression $\dfrac{2}{3} - \dfrac{x-1}{12x+15} + \dfrac{7x^2}{4x^2+5x}$.

Multiplying Rational Expressions

There's no doubt about it—adding fractions is hard, and, to top it off, it really isn't any fun. While it's true that the vast majority of math couldn't easily be classified as "fun," fraction addition leans more toward the other end of the spectrum, best described as "torturously boring" or "monotonous, life-draining, and altogether miserable." Rational multiplication, on the other hand, is pretty easy. And, let's face it, it's about time we got to something easy. This is supposed to be a review chapter for crying out loud! This is supposed to be the easy stuff!

To multiply two fractions, simply multiply the numerators and then do the same with the denominators. So, $\dfrac{a}{b} \cdot \dfrac{x}{y} = \dfrac{a \cdot x}{b \cdot y}$. No common denominators are necessary—just multiply, and you're done (unless, of course, the answer can be simplified).

Example 3: Calculate and simplify the product: $\dfrac{x^3}{x^3+8} \cdot \dfrac{x^2-4x-12}{x}$.

CAUTION

Kelley's Cautions

The fraction $\dfrac{x^3}{x^3+8}$ in Example 3 *cannot* be simplified. Don't be tricked and try to cancel out the x^3 terms $\left(\dfrac{\cancel{x^3}}{\cancel{x^3}+8} \right)$. You can cancel matching pieces of a fraction only if those things either are alone or are *multiplied* by other things.

For instance, if the expression had been $\dfrac{x^3}{x^3 y^2}$, it could have been simplified to $\dfrac{\cancel{x^3}}{\cancel{x^3} y^2} = \dfrac{1}{y^2}$ because x^3 and y^2 are *multiplied*, but because x^3 is *added* in the denominator of the original fraction, you can't cancel it out.

Solution: Begin by factoring the expressions. Note that the denominator of the left fraction is a sum of perfect cubes (whose formula is found in the section "Common Factor Patterns" in Chapter 4).

$$\frac{x^3}{(x+2)(x^2-2x+4)} \cdot \frac{(x-6)(x+2)}{x}$$

Now multiply the numerators together and the denominators together, simplifying the result. Use exponential rules to simplify the x terms $\left(\frac{x^3}{x} = x^{3-1} = x^2 \right)$:

$$\frac{x^3(x-6)\cancel{(x+2)}}{x\cancel{(x+2)}(x^2-2x+4)} = \frac{x^2(x-6)}{x^2-2x+4} = \frac{x^3-6x^2}{x^2-2x+4}$$

You've Got Problems

Problem 3: Calculate the product and simplify: $\dfrac{2x^2-3x}{x^2+6x+9} \cdot \dfrac{4x^3-36x}{2x^2-x-3}$.

Dividing Rational Expressions

When I ask students how to multiply fractions, most take on a robotlike glassy stare and chant this mantra: "Keep, flip, change … keep, flip, change … keep, flip, change." Even though they're eerie in a very *Children of the Corn* sort of way, those enchanted students know what to do: Keep the first fraction as is, flip the second one (take its reciprocal), and change the division sign to a multiplication sign.

In other words, all division problems can be rewritten as multiplication problems because division means the same thing as multiplying by a reciprocal. Think about it in terms of a real number for a second: Dividing a number by 2 means the same thing as multiplying the number by $\frac{1}{2}$ (the reciprocal of 2).

$$10 \div 2 = 5 \qquad 10 \cdot \frac{1}{2} = \frac{10}{2} = 5$$

The "keep, flip, change," or KFC, rule (which comes in original recipe, spicy, or extra crispy) makes sense when you put it in context this way.

Example 4: Calculate the quotient and simplify: $\dfrac{x^2+2x-15}{x-7} \div \dfrac{x^2+9x+20}{x^2-10x+21}$.

Solution: It's always a good idea to factor first when dealing with rational expressions.

$$\frac{(x+5)(x-3)}{x-7} \div \frac{(x+5)(x+4)}{(x-7)(x-3)}$$

Get that spooky, hypnotized look in your eye and repeat after me: "Keep, flip, change."

$$\frac{(x+5)(x-3)}{x-7} \cdot \frac{(x-7)(x-3)}{(x+5)(x+4)}$$

You already reviewed how to multiply fractions, so it's time to set your brain on autopilot.

$$\frac{\cancel{(x+5)}(x-3)\cancel{(x-7)}(x-3)}{\cancel{(x-7)}\cancel{(x+5)}(x+4)} = \frac{x^2-6x+9}{x+4}$$

You've Got Problems

Problem 4: Simplify the expression $\dfrac{a}{b-1} \cdot \dfrac{2a^2-a-1}{a+4} \div \dfrac{2a^2+a}{b^3-1}$. *Hint:* Flip only the fraction immediately following a division sign.

Solving Rational Equations

Have you ever heard the phrase "There's more than one way to skin a cat"? Sure, I know it means "There are multiple solutions to any problem," but I'm not convinced that the anti-cat violence in there is necessary. It's not too far removed from the much less-used children's cliché "There are lots of places to kick your cousin Scott so he won't bruise and get you in trouble with your parents." Honestly, what kind of person is *relieved* to find out that you can remove cat skin in different ways? This phrase needs to be updated.

I offer this replacement candidate for your consideration: "There's more than one way to destroy a denominator." Of course, there are pros and cons to this choice. The pros: Nobody likes fractions, and since a denominator is not a fuzzy animal that purrs when you scratch its head, there are no calls from the ASPCA or PETA. The cons: It's hard to utter this phrase without looking really, *really* geeky, and it could easily trigger a new cliché: "There's more than one place to kick a math geek so we won't try to work math into everyday language anymore."

Nonetheless, there are multiple ways to destroy a denominator. You heard me right: destroy. Once rational expressions appear in an equation, you no longer have to manipulate them—you can actually destroy them completely, leaving a much simpler equation behind. Here are the two best ways to annihilate the denominators in a rational equation:

1. **Rewrite the fractions with common denominators, and then simply drop the denominators altogether.** Once all the denominators in an equation are equal, they're no longer necessary, so it's perfectly legal to eliminate them completely. Therefore, the equation $\frac{2x}{x-1} - \frac{x+3}{x-1} = \frac{7}{x-1}$ has the same solution as the equation $2x - (x + 3) = 7$.

2. **Multiply everything on both sides of the equation by the least common denominator right after you factor the denominators.** This will eliminate the fractions right at the beginning of the problem, which is why I prefer this strategy and use it to solve Example 5, which follows.

One word of warning: Never divide both sides of the equation by a variable, even if you're doing it to try to solve a rational equation, because there's a very real danger that you will actually *eliminate* answers.

Example 5: Solve the equation $\frac{x-5}{x+2} - \frac{2x+3}{x-1} = -1$.

Solution: All of the denominators are unfactorable, so that makes finding the least common denominator a snap: $(x + 2)(x - 1)$. Multiply both sides of the equation by the common denominator to eliminate all the fractions:

Kelley's Cautions

You always have to test the solutions to a rational equation because there's a slim chance that the solution methods I describe here will result in extra or untrue answers. That's a small price to pay for such handy shortcuts, and even with this added step, it saves time in the long run.

$$\frac{(x+2)(x-1)}{1}\left[\frac{x-5}{x+2} - \frac{2x+3}{x-1}\right] = \frac{(x+2)(x-1)}{1}\left[-\frac{1}{1}\right]$$

$$\frac{(x+2)(x-1)(x-5)}{x+2} - \frac{(x+2)(x-1)(2x+3)}{x-1} = \frac{-(x+2)(x-1)}{1}$$

$$(x-1)(x-5)-(x+2)(2x+3) = -(x+2)(x-1)$$

Multiply and combine like terms to find that this is really a simple equation in disguise.

$$(x-1)(x-5)-(x+2)(2x+3)=-(x+2)(x-1)$$
$$x^2-6x+5-\left(2x^2+7x+6\right)=-\left(x^2+x-2\right)$$
$$x^2-6x+5-2x^2-7x-6=-x^2-x+2$$
$$-x^2-13x-1=-x^2-x+2$$
$$-12x=3$$
$$x=-\frac{1}{4}$$

Don't forget to plug your answer back into the original equation to make sure it works (it will).

You've Got Problems

Problem 5: Solve the equation $\dfrac{3}{5}-\dfrac{2}{x-7}=\dfrac{x}{5x-35}$.

Graphing Weird Inequalities

If you haven't read through Chapter 2 yet (especially the sections at the end that discuss graphing inequalities), make sure you do before you try to tackle this section. After all, it's a good idea to get a handle on regular inequalities before you start trying the weird ones.

Ever since I first taught precalculus, I have grouped quadratic and rational inequalities together, in this "weird" category. They are eerily similar because they both require the use of *critical numbers*, which are values that cause an expression either to equal 0 or to become undefined.

Talk the Talk

A **critical number** is a value that causes an expression to equal 0 or become undefined. In a rational expression, critical numbers turn either the numerator or the denominator into 0.

For instance, the expression $\dfrac{x-6}{x+1}$ has two critical numbers: 6 and −1. If you substitute 6 for x, you get $\dfrac{0}{7}$, which equals 0. On the other hand, substituting $x=-1$ results in $\dfrac{-7}{0}$, which is undefined. Essentially, a critical number is a value you can plug in for x that makes an entire expression or just its numerator or denominator equal to 0.

Quadratic Inequalities

There's almost no way to substitute a real number into the x value of a quadratic inequality that would cause that inequality to become undefined. Therefore, when you're searching for its critical numbers, you can focus exclusively on the values that will make the quadratic equal 0. That means you'll need to know how to solve a quadratic equation in order to do these types of problems.

The easiest way to solve a quadratic that's set equal to 0 is to factor it and then set each factor equal to 0. (The quadratic in Example 6 will be factorable so that you can use this approach.) However, if the quadratic were unfactorable, you're not out of luck. You'd just have to use another technique, and I'll give you two alternatives in Chapter 7.

Here's your plan of action for solving quadratic inequalities:

1. **Make sure the quadratic has a 0 on the right side of the inequality sign, and find the critical numbers.** This just means set the polynomial equal to 0 and solve it.

2. **Mark the critical numbers on a number line as either open or closed dots.** Use open dots for < or >, and use closed dots for either ≤ or ≥.

3. **Test the intervals to see which make the original inequality true.** The critical numbers split the number line into pieces. For instance, two critical numbers split a number line into three parts, as illustrated in Figure 5.1.

Figure 5.1

The critical numbers x *and* y *split this number line into three intervals (A, B, and C). In interval notation, they are: A = (–∞, x], B = [x,y), and C = (y,∞). Use brackets to indicate solid dots and parentheses for hollow dots.*

Pick one real number from each interval and plug it into the original inequality. If it makes the inequality true, so will every other value in that interval. The solution for the inequality will be all the valid intervals connected by the word *or*. For instance, if intervals *A* and *C* in

Critical Point

Create the solution graph for a quadratic inequality by darkening the segments of the number line that are solutions for the inequality.

Figure 5.1 satisfied an inequality, the solution would be written like this: "$(-\infty, x]$ or (y, ∞)." Keep in mind that the entire, two-part expression is the answer—not just one or the other, as you might infer from the word *or* in there.

Believe it or not, this technique works not only for quadratic inequalities, but for an inequality of any degree. However, we'll stick to quadratics for now because solving polynomials of higher degree is a more advanced topic and requires the special attention it deserves in Chapter 7.

Example 6: Solve and graph the solution of the inequality $x^2 + 2x > 8$.

Solution: Move all nonzero terms to the left side of the inequality sign.

$$x^2 + 2x - 8 > 0$$

To find the critical numbers of this quadratic inequality, treat it like an equation (change > to =). Factor the quadratic and set each factor equal to 0, to get the critical numbers.

$$(x + 4)(x - 2) = 0$$

$$x + 4 = 0 \quad \text{or} \quad x - 2 = 0$$

$$x = -4 \quad \text{or} \quad x = 2$$

The critical numbers –4 and 2 break the number line into three intervals: $(-\infty, -4)$, $(-4, 2)$, and $(2, \infty)$. Choose one "test value" from each interval—for instance –5, 0, and 3, respectively—and substitute each for x in the original inequality to see which makes the statement true. You'll find that the interval $(-4, 2)$ won't work, but the other two will, so your final answer is "$(-\infty, -4)$ or $(2, \infty)$."

To generate the graph of this solution (shown in Figure 5.2), use open dots at the critical numbers (since the original inequality contains a > sign, not a ≥ sign), and shade in the two intervals you just identified as solutions.

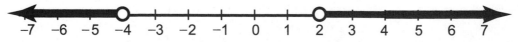

Figure 5.2
Any value on the intervals $(-\infty, -4)$ and $(2, \infty)$ will make the inequality $x^2 + 2x > 8$ true.

You've Got Problems
Problem 6: Solve and graph the inequality $x^2 - 3x \leq 0$.

Rational Inequalities

Solving a rational inequality also requires that you move all nonzero terms to the left side of the inequality sign. However, once they are there, you need to get common denominators and combine them all into one massive fraction. That way, you can set the numerator and denominator of that fraction equal to 0 to quickly identify the critical numbers. Once the critical numbers are identified, it's business as usual.

Example 7: Solve and graph the solution of the inequality $\dfrac{3x}{x-4} \geq -2$.

Solution: Add 2 to both sides of the inequality; then use the least common denominator of $(x-4)$ to add the two terms together:

$$\frac{3x}{x-4} + 2 \geq 0$$

$$\frac{3x}{x-4} + \frac{2}{1} \cdot \frac{x-4}{x-4} \geq 0$$

$$\frac{5x-8}{x-4} \geq 0$$

Now that there's only one fraction on the left side, set the numerator and denominator equal to 0 to find critical values.

$$5x - 8 = 0 \quad \text{or} \quad x - 4 = 0$$

$$x = \frac{8}{5} \quad \text{or} \quad x = 4$$

Just like you did with quadratic inequalities, use test values to get the correct solution: $\left(-\infty, \dfrac{8}{5}\right]$ or $(4, \infty)$; the graph is found in Figure 5.3.

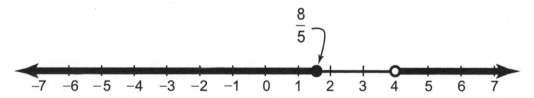

Figure 5.3

Because $\dfrac{8}{5}$ is generated by the numerator, its dot depends on the inequality sign; however, 4 is generated by the denominator of the rational inequality and, therefore, must be marked by an open dot.

Note that you *must* use an open dot to mark any critical numbers that come from the denominator (such as 4, in this case). Those numbers—while still critical—will make the fraction undefined and, therefore, can never be part of the solution. The dot for the other critical number comes (like always) from the inequality sign and whether it allows equality (which, in this problem, it does, so use a closed dot for $x = \frac{8}{5}$).

You've Got Problems

Problem 7: Solve and graph the solution of the inequality $\frac{x+6}{x-2} < \frac{1}{5}$.

The Least You Need to Know

♦ To add or subtract rational expressions, you must have common denominators, preferably the least common denominator of the fractions involved.

♦ Dividing is the same as multiplying by a reciprocal.

♦ You can eliminate the fractions in a rational equation by multiplying the entire equation by the least common denominator (just check your answers when you're finished).

♦ To solve quadratic and rational inequalities, you must use critical numbers, which break the number line into possible solution intervals.

Chapter 6

Functions

In This Chapter

- ◆ Performing function arithmetic
- ◆ Common function graphs to know by heart
- ◆ Graphing functions using transformations
- ◆ Creating inverse functions
- ◆ Identifying asymptotes of rational functions

If you ask a function a question, you know exactly how it will respond. It is that behavior, in fact, that makes it a function: Every input has exactly one output. Ask a function the same question twice or 500 times, and it will always respond the same way. In this chapter, I review the important concepts surrounding functions.

Combining Functions

Just like nearly every other mathematical concept in the universe, functions can be combined using the four basic arithmetic operations: addition, subtraction, multiplication, and division. However, since functions also possess that spiffy input/output behavior that makes them so unique, they can be plugged into one another as well, in a process called the *composition of functions*.

Talk the Talk

A **function** is a mathematical rule such that any input corresponds to exactly one output. This relationship is usually expressed as an equation, like this: $f(x) = x^2 + 3$. In this case, the name of the function is f, the input is x, and the output will be $f(x)$. When one function is plugged into another, it's called the **composition of functions**. To plug the function $g(x)$ into another function, $f(x)$, you would write $f(g(x))$, read "f of g of x." Alternatively, you could write $(f \circ g)(x)$, read "f circle g of x," which means the exact same thing.

Operations on Functions

There's really no trick to performing arithmetic operations on functions, since they're just expressions made up of things you already know how to handle, such as polynomials and radicals. If I asked you to add two fractions, for example, you might not like it, but you could do it. Therefore, you could also add two rational functions. There's absolutely no difference in the procedure just because the expressions are also functions.

Example 1: If $f(x) = 2x^2 - 1$ and $g(x) = \sqrt{x+6}$, evaluate the following:

a. $(f - g)(-3)$

   Solution: Subtract the function $g(x)$ from $f(x)$ and then plug in $x = -3$.

   $$(f - g)(x) = 2x^2 - 1 - \sqrt{x+6}$$
   $$(f - g)(-3) = 2(-3)^2 - 1 - \sqrt{-3+6}$$
   $$(f - g)(-3) = 17 - \sqrt{3}$$

   You could also have calculated $f(-3)$ and $g(-3)$ first ($f(-3) = 17$ and $g(-3) = \sqrt{3}$) and subtracted the results to get the same answer.

b. $(fg)(10)$

   Solution: The function $(fg)(x)$ is the product of $f(x)$ and $g(x)$:

   $(fg)(x) = (2x^2 - 1)(\sqrt{x+6})$. Plug in 10 for x to evaluate $(fg)(10)$:

   $$(fg)(10) = (2(10)^2 - 1)(\sqrt{10+6}) = 199(4) = 796$$

If all you could do to functions was add and subtract them, they'd be a major disappointment. After all, adding expressions, by any other name, is just as boring. However, you've dealt with functions enough to know that, even though they're predictable, they're anything but boring, as you'll see throughout the rest of this chapter.

Composition of Functions

To compose two functions together, you replace the variables in one function with the entire other function—but you have to pay close attention to make sure the right one goes in the right place, because the composition of functions is not commutative. In other words, there's no guarantee that $f(g(x)) = g(f(x))$.

Example 2: If $f(x) = x^2 - 2x$ and $g(x) = \sqrt{x+1}$, evaluate the following expressions:

Critical Point _____

It is *possible* that the composition of functions is commutative, and it's actually an indication of something important. If $f(g(x)) = g(f(x)) = x$ (in other words, when you plug $g(x)$ into $f(x)$ or vice versa, everything cancels out except x), then $f(x)$ and $g(x)$ are called **inverse functions**. More about this comes later in this chapter, in the section "Inverse Functions."

a. $(f \circ g)(x)$

Solution: The expression $(f \circ g)(x)$ means the same as $f(g(x))$. Notice that $g(x)$ is written completely inside of f's parentheses in the expression $f(g(x))$; this tells you to replace every x in $f(x)$ not with a number (as you did in Example 1), but instead with the function $g(x)$.

$$f(x) = x^2 - 2x$$
$$f(g(x)) = \left(\sqrt{x+1}\right)^2 - 2\left(\sqrt{x+1}\right)$$
$$f(g(x)) = x + 1 - 2\sqrt{x+1}$$

How'd You Do That?

A few times in Example 2, a radical expression is eliminated by a square sign, like $\left(\sqrt{x+1}\right)^2 = x+1$. That happens because radicals are actually fractional powers (as you'll remember from the section "Radical Expressions" in Chapter 1).

If I rewrite the radical sign using a $\frac{1}{2}$ power instead, watch what happens: $\left(\sqrt{x+1}\right)^2 = \left[(x+1)^{1/2}\right]^2$. Exponential laws tell you to multiply powers raised to powers, but the exponents $\frac{1}{2}$ and 2 are reciprocals, so their product is 1, leaving a simplified answer of $(x + 1)^1$, or just $x + 1$.

b. $g(f(x))$

Solution: In this expression, you're asked to plug $f(x)$ into $g(x)$, which equals $g(f(x))$.

$$g(x) = \sqrt{x+1}$$
$$g(f(x)) = \sqrt{(x^2 - 2x) + 1}$$

This expression can be simplified! Notice that $x^2 - 2x + 1$ can be factored into $(x - 1)(x - 1)$, or $(x - 1)^2$:

$$g(f(x)) = \sqrt{(x-1)^2}$$
$$g(f(x)) = |x-1|$$

Remember, a quantity to an even power inside a radical with an index that matches that power means you must add absolute value signs.

You've Got Problems

Problem 1: If $h(w) = 2w^2 - 3w - 5$ and $k(w) = w + 1$, evaluate the following expressions:

a. $\left(\dfrac{h}{k}\right)(w)$

b. $(h \circ k)(-1)$

Graphing Functions

Inequality graphs are so simple to draw that it's a little embarrassing. As long as you're careful with your dots, there's really not much to it. Function graphs, on the other hand, are a bit more complex. After all, since functions contain both an *independent variable*, such as x (which holds the input), and a *dependent variable*, such as $f(x)$ (the resulting output), you need to use a graphing system that contains not just one, but two axes. So, good-bye one-dimensional number-line graphs and hello two-dimensional graphs on the coordinate plane.

Talk the Talk

The **independent variable** in a function (usually x) is the one whose value you control because you decide what to plug into it; it is graphed along the horizontal axis. The **dependent variable** (usually $f(x)$ or y) is so named because its value *depends* upon what you plug into the independent variable, and it is graphed along the vertical axis.

Four Important Function Graphs

The vast majority of the graphs you'll generate as a precalculus student can be done quickly and fairly accurately without needing to plot tons of points on the coordinate plane. This is because most graphs are just variations of four basic function graphs, shown in Figure 6.1. Every quadratic function will have a graph that looks a little like $f(x)$ in Figure 6.1, every radical function will look a bit like $j(x)$, every rational will remind you of $k(x)$, and so on.

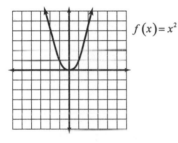
$f(x) = x^2$

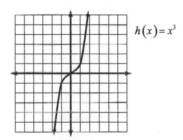
$h(x) = x^3$

Figure 6.1

Memorize the shapes and features of these four basic function graphs, because most of the graphs you'll need to draw in precalculus will actually be a stretched, squished, flipped, or shifted version of one of these.

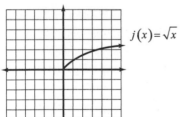
$j(x) = \sqrt{x}$

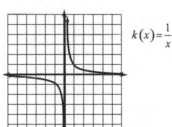
$k(x) = \dfrac{1}{x}$

Even though Figure 6.1 gives you a good idea of what these graphs look like, take a moment and graph each one carefully on graph paper by plotting tons of points. (Plug in a bunch of integer values for x to evaluate $f(x)$, and then plot the points $(x, f(x))$ on the coordinate plane.) That way, you can see why the graphs have the shape they do, and you're not just taking my word for it.

Once you feel comfortable with the graphs and feel that you can draw each one quickly and relatively accurately, you're ready to move on to the next step, function transformation. This will help you draw almost any graph quickly. Remember, the graphs you produce will not be 100 percent accurate, but then again, you're only human, and even graphs based on tons of plotted points are still inherently inaccurate because you still wind up drawing the graphs by hand.

The benefit of the function transformation approach to graphing (and why good instructors teach this method instead of point plotting) is that you actually understand how the numbers in a function affect its graph. You're actually learning important concepts about what makes a graph tick, not just playing Connect the Dots.

Critical Point

You may run across the function $f(x) = [\![x]\!]$ in precalculus and wonder what the heck those double brackets are. They represent the *greatest integer function,* which you've probably never heard of. Like absolute values, these symbols have a very specific purpose, but instead of returning positive numbers (like absolute values do), they return a very specific kind of integer.

To evaluate a greatest integer function, take the number that's inside and figure out what the largest integer is that is *smaller than* (or equal to) that number. For instance, the expression $[\![7.9284]\!]$ is asking "What is the largest integer that is smaller than 7.9284?" The answer is 7.

At first glance, it may look like the greatest integer function just tells you to drop the extra decimal or fraction (like in the expressions $[\![6.246]\!] = 6$ and $\left[\!\left[\frac{21}{5}\right]\!\right] = \left[\!\left[4\frac{1}{5}\right]\!\right] = 4$), and that's true, unless you're dealing with negative numbers.

Notice that $[\![-3.5]\!]$ cannot equal -3 because $-3 > -3.5$, and the greatest integer function demands that the answer be less than what's inside those double brackets. Therefore, when you're finding the greatest integer function value of a negative number, the answer is exactly one less than what you'd get if you'd simply chopped off the decimals: $[\![-3.5]\!] = -4$ and $\left[\!\left[-9\frac{2}{3}\right]\!\right] = -10$.

Function Transformations

When I was young, I owned a fairly good-size collection of toys called Transformers. You've probably heard of them, played with them, bought them for your kids, or inadvertently gotten one lodged in your trachea. In any case (except arguably the last), they were great fun. With just a few simple twists and clicks, you could turn a plastic jet into a cool action figure with anatomy that slightly less resembled a jet. Like the good television-fed generation we were, my brother Dave and I would wage epic Transformer battles, reciting the catch phrases fed to us by the commercials: "Transformers! More than meets the eye! Transformers! Robots in disguise!"

We learned a valuable lesson from those toys as the "Autobots waged their battles to destroy the evil forces of the Decepticons" (does anyone besides me also have that theme song etched into his brain?): A few simple tweaks to a well-known concept can produce a result that, while unique, still looks uncannily like what you started with. ("Hey, why does that giant robot have a tape player built into his chest?")

The same thing goes for algebra—a negative sign here or a constant jammed in there can take one of the graphs I introduced in the previous section (the four basic function graphs of Figure 6.1) and create approximately 70 percent of the graphs you'll need to draw as a precalculus student. Here are the details of those tweaks and how they affect the graph of a generic function $f(x)$:

- **A constant added to or subtracted from the function causes a vertical shift in the graph.** In other words, the graph of a function $f(x) + 2$ is created by moving the entire graph of $f(x)$ up two units. Similarly, you would move the original graph of $f(x)$ down three units to get the graph of $f(x) - 3$.

- **A constant added to or subtracted from the input of a function shifts its graphs horizontally.** The difference between this and the previous transformation is the location of the constant that's added or subtracted. In this case, it appears *inside the function itself*, like $f(x + 2)$ instead of $f(x) + 2$.

  Adding within a function causes the graph to move that many units *left*, and subtracting moves the graph *right*, which is exactly the opposite of what you'd think. So, the graph of $f(x + 5)$ looks like the graph of $f(x)$ moved five units left, and the graph of $f(x - 1)$ is the graph of f moved one unit to the right.

> **Critical Point** _____
>
> Notice that transformations applied to the input of a function have the opposite effect than you probably expect. Adding to the input moves the function *left*, not right; multiplying the input by a large number actually causes the graph to *shrink* horizontally.

- **Multiplying a function by a constant stretches or squishes the function vertically.** Simply stated, the graph of $5f(x)$ stretches five times as high and low as the graph of *f(x)*. Multiplying by a number greater than 1 stretches a function away from the *x*-axis, but multiplying by a number between 0 and 1 actually squishes the function toward the *x*-axis.

Notice, however, that this does not change a function's *x*-intercepts—because those points technically have a height of 0, five times that height still equals 0.

♦ **Multiplying the input of a function by a constant stretches or squishes the graph horizontally.** In this kind of transformation, the multiplication occurs within the function itself, such as *f(3x)* rather than *3f(x)*. Multiplying by a number greater than 1 causes the graph to become narrower, jammed closer to the origin (*f(3x)* has a graph three times as narrow as *f(x)*), whereas multiplying by a number between 0 and 1 actually stretches the graph, according to the reciprocal of that number (the graph of $f\left(\frac{1}{2}x\right)$ is twice as wide as the graph of *f(x)* because $\frac{1}{2}$ is the reciprocal of 2).

♦ **Multiplying a function by a negative number causes its graph to reflect over the *x*-axis.** This doesn't change the horizontal location of the points; it only moves them an equal distance across the *x*-axis. In other words, every point (*x,y*) on the graph turns into (*x,−y*), flipping the graph over the *x*-axis like in Figure 6.2.

Figure 6.2

The graphs of f(x) and −f(x) are reflections of one another across the x-axis, while the graphs of f(x) and f(−x) are reflections across the y-axis.

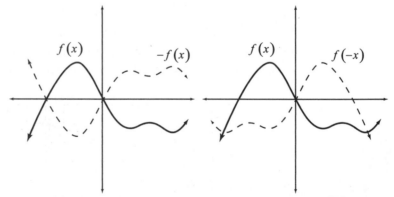

Reflection across the *x*-axis   Reflection across the *y*-axis

♦ **Multiplying a function's entire input by a negative number reflects its graph across the *y*-axis.** This time, the height of the points does not change, but they are moved an equal distance across the *y*-axis. In other words, every point (*x,y*) will turn into (*−x,y*), flipping the graph across the *y*-axis like in Figure 6.2.

♦ **Taking the absolute value of a function moves all the points on its graph above the x-axis.** Any piece of f(x)'s graph that was already above the x-axis will not change, but any piece that dipped below it will be reflected across the x-axis in the graph of $\left|f\left(x\right)\right|$.

♦ **Taking the absolute value of a function's input causes the left side of the graph to clone the right side.** To graph $f\left(\left|x\right|\right)$, begin by drawing the regular graph of f(x). When you're finished, erase any part of the graph to the left of the y-axis. Leave the right part of the graph alone, but reflect a copy of it across the y-axis to complete the graph. This is the most complex of the graph transformations, and because it is so bizarre, I feature it in Example 3.

If there are multiple transformations in a given graph, do them in this order: reflections, stretches, absolute values, and then shifts.

Example 3: Graph the function $g\left(x\right)=\sqrt{\left|x\right|}+3$.

Solution: Start with the untransformed graph of $\sqrt{x}$, which you should have memorized. There are two transformations in g(x), absolute value signs around the input and 3 added to the function. You should apply the absolute value transformation first, so erase any portion of the graph to the left of the y-axis. (In this case, however, no part of the graph exists to the left of the y-axis.) Now reflect a copy of what's left across the y-axis, like you would in the transformation f(–x). You end up with a graph that looks like a seagull in flight.

Finally, shift the entire graph up three units so the seagull's body is located at the point (0,3) instead of (0,0). You'll end up with the final graph in Figure 6.3.

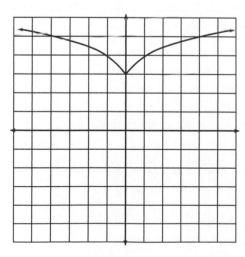

Figure 6.3

The graph of $g\left(x\right)=\sqrt{\left|x\right|}+3$ is created by cloning the graph of $\sqrt{x}$ across the y-axis and then moving the result up three units.

Problem 2: Graph the function $h(x) = -(x + 3)^2 - 1$.

Inverse Functions

Earlier in the chapter, during my explanation of the composition of functions, I mentioned a special relationship between two functions that canceled each other out when they were plugged into one another. If $f(g(x)) = g(f(x)) = x$, then $f(x)$ and $g(x)$ are inverse functions and can be written as $g(x) = f^{-1}(x)$ or $f(x) = g^{-1}(x)$.

Kelley's Cautions

Inverse function notation, $f^{-1}(x)$, does *not* mean that $f(x)$ is raised to the −1 power; that would be written like this: $(f(x))^{-1}$ and means something else entirely. Whereas $f^{-1}(x)$ means "the inverse of $f(x)$" or "f inverse of x," $(f(x))^{-1}$ means "the reciprocal of $f(x)$."

You already know a bunch of inverse functions, even if you don't know you know them (trust me, I know). Every time you solve an equation, you're using inverse functions. For example, if asked to solve $x^2 = 4$, the best strategy would be to take the square root of both sides to get $\sqrt{x^2} = \sqrt{4}$, which means $x = \pm 2$. When you plugged the function x^2 into the square root function, they canceled one another out, leaving only x on the left side of the equation and allowing you to solve the problem.

Inverse Function Behavior

The relationship between a function and its inverse has very specific characteristics, including these important ones:

♦ **An inverse function reverses the ordered pair of the original function.** Let's say you plug 5 into the function $f(x)$ and get out −2, so $f(5) = -2$. The inverse function reverses those numbers; the input becomes the output and vice versa. Therefore, $f^{-1}(-2) = 5$. This is arguably the most important feature of inverse functions and is what makes them cancel out one another. In this case, we have $f(f^{-1}(-2)) = f(5) = -2$ and $f^{-1}(f(5)) = f^{-1}(-2) = 5$.

♦ **Inverse function graphs are reflections of one another across the line $y = x$.** The coordinate-reversing property of inverse functions I described in the previous bullet has a peculiar effect on their graphs: They mirror their original functions across the line $y = x$ (the simple line passing through the origin with slope 1), as demonstrated in Figure 6.4.

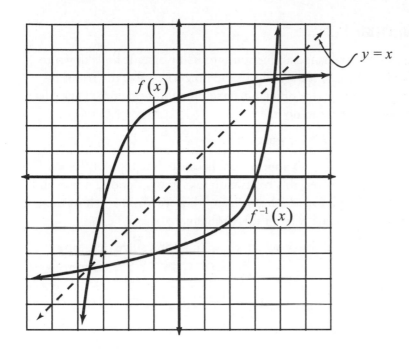

Figure 6.4

Just as the graphs of f(x) and –f(x) are reflections of one another across the x-axis, the graphs of a function and its inverse are reflections of one another across the dotted line y = x.

◆ **A function must pass the horizontal line test to possess a valid inverse.** Remember the vertical line test? It said that if any vertical line drawn through a graph hit in more than one spot, the graph couldn't represent a function. Well, if any horizontal line strikes the graph of a function more than once, that function cannot have an inverse.

How'd You Do That?

Are you wondering why a function must pass the horizontal line test to qualify for an inverse? Let's say that some function $f(x)$ fails the horizontal line test because both $f(1) = 7$ and $f(5) = 7$. (In other words, the horizontal line $y = 7$ will hit the graph of $f(x)$ twice, once when $x = 1$ and once when $x = 5$.)

Remember, an inverse function reverses ordered pairs, so $f^{-1}(7) = 1$ and $f^{-1}(7) = 5$. Hold on a minute—that's not allowed! If $f^{-1}(x)$ is truly a function, it must have exactly one output for every input. In the case of $f^{-1}(x)$, an input of 7 would output either 1 or 5. Failing the horizontal line test guarantees that a function's inverse is not a function at all.

Creating an Inverse Function

You can use a simple process to change a function into its inverse. It boils down to this: Replace the $f(x)$ with y, switch the x and y in the function, solve the result for y, and, when finished, replace that y with official inverse function notation, $f^{-1}(x)$. Here's a quick example to make sure you can do it.

Example 4: If $f(x) = x^2 - 2$ (when $x \geq 0$), what is $f^{-1}(x)$?

Solution: The restriction ($x \geq 0$) has to appear here because the graph of $f(x) = x^2 - 2$ fails the horizontal line test. However, this restriction tells you to ignore, erase, and forget about the part of the graph for which $x < 0$. In other words, the half of the graph to the left of the y-axis doesn't exist in this problem.

Replace $f(x)$ with y to get $y = x^2 - 2$. Now switch the x and y and solve that for y.

$$x = y^2 - 2$$
$$x + 2 = y^2$$
$$\sqrt{y^2} = \pm\sqrt{x+2}$$

The final answer is $f^{-1}(x) = \sqrt{x+2}$, but not $f^{-1}(x) = -\sqrt{x+2}$. Why? Only the positive radical has a graph that's a reflection of $f(x)$ across the line $y = x$. (The graph of $f^{-1}(x) = -\sqrt{x+2}$ reflects the piece of $f(x)$ that was eliminated by the restriction $x \geq 0$, so that's not included in the answer.)

You've Got Problems

Problem 3: If $g(x) = 2x - 6$, what is $g^{-1}(x)$?

Asymptotes of Rational Functions

Think back to the graph of $f(x) = \frac{1}{x}$ for a moment; what made it different from the other function graphs you memorized in this chapter? It was the only one that had a break in it—it's impossible to draw the graph from one end to the other without lifting your pencil at least once.

The function has to break because $f(0)$ is undefined; the fraction $\frac{1}{0}$ doesn't have a real number value, so the graph contains a vertical *asymptote* whose equation is $x = 0$. An asymptote is a line that, while not actually part of the graph, shapes it nonetheless because the graph actually bends to avoid touching that line. Think of an asymptote line as a boundary that graphs will get nearer and nearer to for all of infinity but must never touch (probably because it'll make them all itchy, and being itchy for all of eternity is not something graphs enjoy).

Even the simplest of rational functions usually has at least one asymptote on its graph, but there are three different kinds of asymptotes it may possess. Here's how to figure out what those asymptotes are:

Talk the Talk

An **asymptote** is a line representing unattainable points for the graph of a rational function. It is usually drawn as a dotted line on the coordinate plane but is not technically part of the graph.

◆ **Vertical asymptotes.** To find the vertical asymptotes of a function, factor it completely and then set each factor of the denominator equal to 0. Every solution you get that does not make the numerator 0 will give you a vertical asymptote of the function.

Critical Point

Horizontal asymptotes, like all horizontal lines, have equations that start with $y =$. Vertical asymptote equations begin with $x =$. Slant asymptotes are linear equations in slope-intercept form, so they look like $y = mx + b$, with numbers instead of m and b, of course.

◆ **Horizontal asymptotes.** If the degree of the numerator is n and d is the degree of the denominator, you can compare the values of n and d to figure out the horizontal asymptotes:

  ◆ If $n > d$, there are no horizontal asymptotes.

  ◆ If $n < d$, the function has exactly one horizontal asymptote: $y = 0$.

  ◆ If $n = d$, the function has one horizontal asymptote: $y = \dfrac{a}{b}$, where a is the leading coefficient of the numerator and b is the leading coefficient of the denominator.

◆ **Slant asymptotes.** Any asymptote that is neither vertical nor horizontal is described as a *slant asymptote*, but these occur only in one specific case: When the numerator's degree is exactly one more than the denominator's. Use long or synthetic division to divide the numerator by the denominator, and ignore the remainder. Just set y equal to the quotient you end up with; that's the equation of the slant asymptote.

A rational function can have a bunch of vertical asymptotes, but it can have only one horizontal asymptote or one slant asymptote.

Example 5: Determine how many asymptotes the function $f(x) = \dfrac{2x^2 - 9x + 9}{x - 4}$ has, and give the equation of each.

Solution: Start by factoring the numerator and denominator.

$$f(x) = \frac{(2x - 3)(x - 3)}{x - 4}$$

Now you should check for vertical asymptotes. Set the denominator equal to 0, and you'll get $x = 4$. Because this value does not cause either factor in the numerator (and, therefore, the entire numerator) to equal 0, $x = 4$ is a vertical asymptote of the function.

Are there horizontal asymptotes? The degree of the numerator (2) is greater than the degree of the denominator (1), so, no, there aren't any. However, that one degree difference is a tip-off that there will be a slant asymptote.

To calculate that slant asymptote, divide the original numerator by the denominator: $(2x^2 - 9x + 9) \div (x - 4)$. Because the divisor is a linear binomial, you can use synthetic division. Don't forget that the opposite of the binomial's constant always goes in the box. (Remember, you subtracted in long division).

$$\begin{array}{r|rrr} 4 & 2 & -9 & 9 \\ & & 8 & -4 \\ \hline & 2 & -1 & 5 \end{array}$$

The quotient is $2x - 1$ (ignore the remainder of 5), so the equation of the slant asymptote is $y = 2x - 1$. Check out Figure 6.5 to see how these two asymptotes act as boundaries and squeeze the graph of $f(x)$ into a restricted space.

Figure 6.5

Notice how the graph of
$f(x) = \dfrac{2x^2 - 9x + 9}{x - 4}$ *bends to*
avoid the asymptotes you
found in Example 5, almost
as if they were infected with
a raging case of cooties.

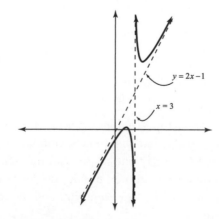

Problem 4: Determine how many asymptotes the function $g(x) = \dfrac{3x^2 - 11x - 4}{x^2 - 12x + 32}$ has, and give the equation of each.

The Least You Need to Know

♦ The expression $(f \circ g)(x)$ means $f(g(x))$—plug the entire function $g(x)$ in for the independent variable of $f(x)$.

♦ Most function graphs can be created by applying simple transformations to one of the following graphs: $y = x^2$, $y = x^3$, $y = \sqrt{x}$, or $y = \dfrac{1}{x}$.

♦ If a function's graph contains the point (a,b), the graph of its inverse will contain the point (b,a).

♦ Rational functions may be shaped by vertical, horizontal, and slant asymptotes.

Part 2 Nonlinear Equations and Functions

If you understood most of the review chapters (even if some of it was a surprise), you probably feel pretty confident right now. If, however, the review section was pretty tricky and it took you a while to rub off your algebraic rust to work through the problems, you may be wondering, "How am I ever going to make it through the rest of this book without throwing myself off of a cliff?" Surprisingly, both camps of readers (the confident and the freaked out) will find solace in the chapters to come. Even though the material from here to the end of the book will likely be new and challenging, the pace will also slow down a bit.

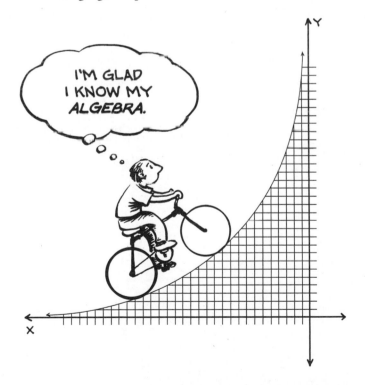

Chapter 7

High-Powered Functions and Equations

In This Chapter

- ◆ Cracking unfactorable quadratic equations
- ◆ Drawing conclusions based on the leading coefficient and rational root tests
- ◆ Applying Descartes' Rule of Signs
- ◆ Finding roots of functions that aren't quadratic

As you mature in algebra, you realize that there's not always one answer to a problem. As you graduate from linear equations and move into quadratic and other equations with even higher exponents, the number of possible answers increases, and the solution methods you use to reach those answers get a little bit more difficult. That's okay because it means you're growing up in math and can handle a little uncertainty. In this chapter, you'll not only come to terms with equations that possess multiple solutions, but you'll also learn how to use the clues embedded in those equations to figure out exactly what the solutions are.

Solving Quadratic Equations

Toward the end of Chapter 5, I reviewed the following as the simplest way to solve a quadratic equation:

1. Set the quadratic equal to 0.

2. Factor the polynomial.

3. Set each factor equal to 0 and solve.

Kelley's Cautions

The factoring technique for solving quadratics is guaranteed to work only if you set the polynomial equal to 0 *before* you factor it.

Therefore, to solve the quadratic equation $x^2 - 9x + 6 = -14$, you first add 14 to both sides to set the quadratic equal to 0: $x^2 - 9x + 20 = 0$. You then factor the left side of the equation, set each factor equal to 0, and solve the resulting mini-equations.

$$(x-4)(x-5) = 0$$
$$x - 4 = 0 \quad \text{or} \quad x - 5 = 0$$
$$x = 4 \quad \text{or} \quad x = 5$$

This is so simple and straightforward that you may wonder why on earth you'd ever want to learn another method for solving quadratic equations. The reason is this: Not all quadratics will factor nicely into a pair of binomials. In fact, most of them won't. Luckily, you've got two alternative solution methods (completing the square and the quadratic formula) that solve any quadratic you could possibly come up with.

Most students gravitate toward the quadratic formula because it's a simple "plug and chug" process—memorize the formula, plug in the numbers, simplify the equation, bada bing, bada boom, game over. However, you should definitely learn how to complete the square as well, because you'll need to know how when you get to conic sections in Chapters 15 and 16.

Completing the Square

Your mathematical goal when completing the square is to rewrite the quadratic so that it's not only factorable, but actually a perfect square. It may sound tricky, but there is a very specific procedure to follow for every problem of this type, which I illustrate in the following example.

Example 1: Solve the equation $2x^2 - 12x + 4 = 0$ by completing the square.

Solution: Danger! You can never complete the square unless the leading coefficient (the coefficient of x^2) is 1, but that is easily remedied here—just divide everything by the leading coefficient of 2. (As long as you divide an entire equation by a constant, the solutions are unaffected.)

$$x^2 - 6x + 2 = 0$$

Now move the constant to the right side of the equation. This is different than the factoring and quadratic formula techniques, which both require an equation set equal to 0.

$$x^2 - 6x = -2$$

Here comes the key step to correctly completing the square. Multiply the coefficient of the x term by $\frac{1}{2}$: $\frac{-6}{2} = -3$. Take that result (-3) and square it: $(-3)^2 = 9$. Add this value to both sides of the equation.

$$x^2 - 6x + 9 = -2 + 9$$

If you did everything right, the quadratic on the left side of the equation should factor into two matching binomials, in this case $(x - 3)$ $(x - 3)$, which can be written as a squared term: $(x - 3)^2$.

$$(x - 3)^2 = 7$$

Critical Point

When you take half of the x coefficient (on your way to squaring the result and adding it to both sides), make a special note of the result (in this case, -3). This will be the number that appears next to x when you finally rewrite the quadratic as a square: $(x - 3)^2 = 7$. That makes factoring even the toughest perfect squares a snap.

Solve this equation by taking the square root of both sides and isolating x:

$$\sqrt{(x-3)^2} = \pm\sqrt{7}$$
$$x - 3 = \pm\sqrt{7}$$
$$x = 3 \pm \sqrt{7}$$

So, there are two solutions to the equation $2x^2 - 12x + 4 = 0$: $x = 3 + \sqrt{7}$ and $x = 3 - \sqrt{7}$.

The Quadratic Formula

The *quadratic formula* allows you to find the solutions to the quadratic equation $ax^2 + bx + c = 0$ by plugging its coefficients a, b, and c into this formula:

$$x = \frac{-b \pm \sqrt{b^2 - 4ac}}{2a}$$

Once you've got the formula memorized (and you should do that right away, if you haven't already), all you have to do is make sure that the quadratic you're trying to solve is set equal to 0, stick the coefficients in the right places, and simplify.

The quadratic formula is not magic. Someone didn't just happen to stumble across it one day. It's actually what you get when you solve the equation $ax^2 + bx + c = 0$ by completing the square. The process is actually pretty neat, so let me show you how it works.

Remember, the leading coefficient of $ax^2 + bx + c = 0$ must equal 1, so divide everything by a, like you divided by 2 in Example 1 (and move the constant to the right).

$$x^2 + \frac{b}{a}x = -\frac{c}{a}$$

Multiply the middle term by $\frac{1}{2}$ to get $\frac{b}{2a}$ and square the result: $\left(\frac{b}{2a}\right)^2 = \frac{b^2}{4a^2}$. That fraction should be added to both sides of the equation:

$$x^2 + \frac{b}{a}x + \frac{b^2}{4a^2} = -\frac{c}{a} + \frac{b^2}{4a^2}$$

Factor the left side and add the fractions on the right, making sure to use common denominators.

$$\left(x + \frac{b}{2a}\right)^2 = \frac{-4ac + b^2}{4a^2}$$

Solve for x, and you'll end up with the quadratic formula.

$$\sqrt{\left(x + \frac{b}{2a}\right)^2} = \pm\sqrt{\frac{-4ac + b^2}{4a^2}}$$

$$x + \frac{b}{2a} = \pm\frac{\sqrt{b^2 - 4ac}}{2a}$$

$$x = \frac{-b \pm \sqrt{b^2 - 4ac}}{2a}$$

That's a little tough, so don't panic if you don't fully understand it. I just thought you might be a bit curious to see that you have all the math knowledge necessary to generate this formula yourself.

Example 2: Solve the equation from Example 1 ($2x^2 - 12x + 4 = 0$) again, this time using the quadratic formula.

Solution: The equation is already in the form $ax^2 + bx + c = 0$ (because it's set equal to 0), so plug $a = 2$, $b = -12$, and $c = 4$ into the quadratic formula and simplify. (In Example 1, you divided everything by 2 to get $x^2 - 6x + 2 = 0$, and I told you it wouldn't affect the answers. Therefore, you could also plug those coefficients ($a = 1$, $b = -6$, and $c = 2$) into the quadratic formula, and you'll get the same answer.)

$$x = \frac{-b \pm \sqrt{b^2 - 4ac}}{2a}$$

$$x = \frac{-(-12) \pm \sqrt{(-12)^2 - 4(2)(4)}}{2(2)}$$

$$x = \frac{12 \pm \sqrt{112}}{4}$$

$$x = \frac{12}{4} \pm \frac{4\sqrt{7}}{4}$$

$$x = 3 \pm \sqrt{7}$$

This is the same as the answer obtained by completing the square in Example 1.

You've Got Problems

Problem 1: Solve the equation $3x^2 - 15 = 6x$ twice, once by completing the square and once via the quadratic formula. You should get the same answer(s).

Finding Roots Forensically

As you probably know if you've watched the show *C.S.I.* or one of its numerous spin-offs, such as *C.S.I. Miami* or *C.S.I. Clown College*, forensic analysis is the study of teeny, microscopic clues at the scene of a crime. Because a killer (usually) isn't just going to step forward and confess to a crime, homicide detectives (following the lead of historically successful crime-solvers like Fred, Daphne, Velma, Shaggy, and, of course, Scooby Doo) must look for clues.

In the land of Scooby Doo, these clues are pretty obvious—rubber werewolf masks, antiquated movie projectors used to simulate ghosts, and so on—but in real life, clues are much more subtle. A hair, fiber, footprint, or drop of blood (although seemingly small and insignificant) can give the investigation direction and ultimately provide evidence leading to a conviction.

Calculating solutions to equations is not easy once the degrees are larger than 2. There are no straightforward, easily memorized formulas like the quadratic formula that instantly give you the solutions of a cubic polynomial, let alone a quartic, quintic, or what have you. You'll have to conduct your own crime scene investigation (*C.S.I. Math Geek*) to get some background on the equation and identify likely solution suspects. For a big chunk of this chapter, I'll train you to comb an equation crime scene and introduce you to the tools that will help you nab the answers to complex equations.

Equations vs. Functions

I need to adjust your thinking for a minute. I need you to visualize equations in a slightly different way: as functions. For all intents and purposes, solving a polynomial equation (which is our focus for the next few pages) is the same thing as calculating the *roots* of a polynomial, the x values at which that function will have a value of 0.

Talk the Talk

The **roots** of a function are the numbers that make a function equal 0; therefore, if c is a root of the function $g(x)$, then $g(c) = 0$. (Because of this, the roots are also called the **zeros** of a function.) Graphically speaking, a function's roots are the x-intercepts of the graph. That makes sense because if c is a root of $g(x)$ (and, therefore, $g(c) = 0$), the x-axis point $(c,0)$ must be on the graph of $g(x)$.

To change an equation into a function, simply solve it for 0 and replace the 0 with function notation, as in $f(x)$ or $g(x)$. For instance, the solutions to the equation $x^3 + 2x = 5x - 1$ will also be the roots of the function $f(x) = x^3 + 2x - 5x + 1$. As you can see, nothing drastically changed there. All the terms are the same; you basically just moved them all to one side of the equation and stuck a function label on the other side. Now that you're dealing with a function, though, you can apply all the forensic tests I'm about to introduce.

The Fundamental Theorem of Algebra

When trying to calculate the roots of a function, it sure would be nice, at the very least, to know how many you're actually looking for. That answer is provided by the *Fundamental Theorem of Algebra*, which states that a polynomial function with degree n will always have exactly n roots. In other words, a polynomial whose highest degree is 3 will have exactly three roots.

Even though the Fundamental Theorem is handy, there are two tricky things about implementing it that you need to keep in mind:

- **The roots may repeat.** Unfortunately, the roots aren't necessarily unique. It couldn't be simpler to find the roots of $f(x) = (x + 1)(x + 1)(x - 7)$. Since it's factored, all you have to do is set each factor equal to 0 and solve, like you did to solve simple quadratic equations. You'll get roots of –1, –1, and 7. Technically, this equation has three roots, even though two of them are exactly the same. In this problem, –1 is called a *double root* because it occurs twice in the solution. (In case you're wondering, you have to write the answer only once.)

- **The roots may not be pretty.** A function may have irrational or complex roots, which are not as easy to calculate as rational roots. However, once you identify those rational roots, the uglier roots are much easier to calculate, as you'll see in Example 5b at the end of the chapter.

Now that you know how many roots a function may have, it's time to start figuring out what they are.

Leading Coefficient Test

Although it is not extraordinarily handy, the leading coefficient test at least gives some idea of a function's graph. Specifically, it tells you what the far right and left ends of the function do as they speed off toward infinity, what mathematicians call the "end behavior" of the function. All you need for this test is the degree of the function and (as you may have guessed, based on its name) the function's leading coefficient. Here are the conclusions you can draw:

- **The ends of a function with an even degree point in the same direction.** If the leading coefficient is positive, both ends point upward. (Picture in your mind the graph of $f(x) = x^2$, which has a degree of 2 and a positive leading coefficient; the U-shape, parabolic graph points up at both ends.) On the other hand, if the leading coefficient is negative, both ends of the graph point down.

◆ **The ends of a function with an odd degree point in opposite directions.** If the leading coefficient is positive, the left end points down and the right end points up (like the graph of $g(x) = x^3$). Alternatively, a negative leading coefficient indicates an upward-pointing left end and a right end that heads down.

In case you forget these rules of thumb, there is another (only slightly longer) way to find the end behavior of a function: Plug a really gigantic number (such as 5,000) and a really huge negative number (such as –5,000) into the function and see if your result is positive or negative. For instance, if $f(-5,000) = 729,300$, then clearly the left end of $f(x)$ is heading upward.

Example 3: Describe the end behavior of $j(y) = 6 + y^2 - 5y^3$.

Solution: The degree of $j(y)$ is odd (3), so the ends of $j(y)$ head in opposite directions. Because the leading coefficient is negative (–5), you can conclude that the left end goes up and the right end of the graph goes down.

Descartes' Rule of Signs

You use Descartes' Rule of Signs to intelligently guess how many positive and negative roots a function will have. Unfortunately, it's usually just a guess, and you won't always get definitive answers. Instead of a more promising result like "$f(x)$ will have one positive and two negative roots," Descartes' Rule of Signs will make this sort of conclusion: "$f(x)$ will have either five, three, or one positive root(s) and either four, two, or zero negative roots."

Believe it or not, that is often extremely useful information, as you'll see in Example 4. Here's how Descartes' Rule of Signs works:

1. **Put the function in standard form.** The terms need to be in order, from the highest exponent to the lowest, or the test won't work.

2. **Count the sign changes.** Start with the first term and go one by one to the last term, counting the number of times the terms change sign. For example, the function $f(x) = -2x^3 - x^2 + 3x - 9$ has two sign changes (they occur after the second term, changing from negative to positive, and after the third term, changing back to negative). That number (in this case, 2) is a possible number of positive roots for the function.

Critical Point

Descartes' Rule of Signs is not like long or synthetic division. You don't have to worry if powers are missing when you count sign changes. You don't have to rewrite $f(x) = x^5 + 2x - 1$ as $x^5 + 0x^4 + 0x^3 + 0x^2 + 2x - 1$ because the missing coefficients are all 0, which is neither positive nor negative, so they never count as a sign change.

3. **Subtract multiples of 2.** Take the number of sign changes you got from the last step and subtract multiples of 2 until you get a negative number. All the non-negative results (including 0) are possible positive root counts. For instance, if step 2 gives you seven sign changes for a function $m(x)$, then $m(x)$ must have seven, five, three, or one positive root(s). If a function $b(y)$ has six sign changes, it must have six, four, two, or zero positive roots.

4. **Find the negative roots.** Go back to the original function and substitute $-x$ for x. Count the sign changes again and subtract multiples of 2 to calculate the number of possible negative roots.

As weird as this technique is with all its subtracting of 2's and ambiguous results, it delivers an impressive amount of information for the small amount of work required. Keep in mind that this test gives you information only about a function's real number roots, not complex ones. That's because complex numbers aren't "positive" and "negative" in the same sense real numbers are, so they are excluded from the test.

Example 4: Apply Descartes' Rule of Signs to the function $h(x) = 4x^5 - x^2 + 2x - 11$.

Solution: The function is already in standard form, so you can immediately count the sign changes; there are three because the signs change between each of the four terms. Therefore, the function has either three positive roots or one positive root. (Don't forget to subtract 2's from the sign change count to get other possible answers until subtracting 2 doesn't make sense anymore. Saying there are "–1 possible positive roots," for instance, is just crazy talk.)

Now evaluate $h(-x)$ by plugging in $-x$ for every x in the function and simplifying:

$$h(-x) = 4(-x)^5 - (-x)^2 + 2(-x) - 11$$
$$h(-x) = -4x^5 - x^2 - 2x - 11$$

Notice that every term containing an odd exponent has a different sign than it did in the original function $h(x)$. The total number of sign changes here is zero—everything is negative. Therefore, $h(x)$ has no negative roots. Do you see why Descartes' Rule of Signs is so useful in this problem? According to the Fundamental Theorem of

Algebra, $h(x)$ must have five roots (since it has degree 5), but it has only three or one real root(s), which are positive. You can conclude that the other roots (there will be either two or four of them, depending upon the actual number of positive roots) are complex.

Rational Root Test

Let's review. You know how many roots a polynomial function must have (the Fundamental Theorem of Algebra), where the edges of its graph will go (the leading coefficient test), and approximately how many positive and negative real roots the function will have (Descartes' Rule of Signs). That's all well and good, but are you wondering when we're actually going to find out what those roots are? That time is now.

Talk the Talk

The **rational root test** guarantees that a function with leading coefficient a and constant c will have rational roots of the form $\pm\frac{y}{x}$, where y is a factor of c and x is a factor of a.

The *rational root test* gives you a long list of *possible* rational roots for a function. The key word here is *possible*. By the end of this test, you'll end up generating a long list of integers and fractions, any of which *could* be roots of the function. Unfortunately, the rational root test doesn't actually tell you which ones *are* the roots; you'll have to check that yourself. Luckily, it's not too hard to do; in fact, you use a technique that I reviewed with you already, way back in Chapter 3.

Here's how you apply the rational root test:

1. **Identify the leading coefficient and constant in the function.** For reference, I'll refer to the leading coefficient as a and the constant as c. Ignore the signs of both a and c—whether the function has a leading coefficient of 4 or –4 (or a constant of 10 or –10) doesn't matter for this test; only the number itself matters.

2. **List all the factors of a and c separately.** Note that you're not creating the prime factorization; you're just listing all the numbers that divide into a and c evenly. For instance, although 12 has a prime factorization of $2^2 \cdot 3$, its list of factors is 1, 2, 3, 4, 6, and 12.

3. **Write every possible combination of c's factors divided by a's factors.** In other words, list every possible fraction in which the numerator is a factor of c and the denominator is a factor of a. The easiest way to do this is to take the first factor of c and use each factor of a as a different denominator. Then use the second factor of c, and so on.

For example, let's say $c = 4$ and $a = 5$. The factors of c are 1, 2, and 4, and the factors of a are 1 and 5. Divide the first factor of c by each factor of a: $\frac{1}{1}$ and $\frac{1}{5}$. Now do the same thing with the second factor of c $\left(\frac{2}{1} \text{ and } \frac{2}{5}\right)$ and the last factor of c $\left(\frac{4}{1} \text{ and } \frac{4}{5}\right)$. You end up with this final list of rational numbers: $\frac{1}{5}, \frac{2}{5}, \frac{4}{5}, 1, 2$, and 4.

4. **Include the opposite of every item in the list you just generated to get the final list of possible roots.** This doubles the number of rational numbers you got in step 3 by tacking on the opposite of each. So, the final list of possible rational roots for the function is this: $-4, -2, -1, -\frac{4}{5}, -\frac{2}{5}, -\frac{1}{5}, \frac{1}{5}, \frac{2}{5}, \frac{4}{5}, 1, 2$, and 4.

Any of those 12 numbers could be roots to *any* function whose leading coefficient is 5 and whose constant is 4. Therefore, this is the possible list of roots for the function $f(x) = 5x^3 + 9x + 4$, as well as the function $g(y) = -5x^6 + 4x^5 - 11x^2 + 2x - 4$. (Remember, in the rational root test, the signs of a and c are irrelevant, thanks to step 4, which makes both the positive and negative versions of every root a possibility.)

You've Got Problems

Problem 2: Identify all possible rational roots of the function $h(x) = 7x^2 - 6x^3 + 2$.

Wrapping Up the Case

Although the rational root test identifies a list of potential suspects, you must figure out which of those suspects are actually roots. Before you get started, though, there's one thing I want you to remember: If r is a root of the function $f(x)$, then $(x - r)$ will be a factor of $f(x)$, and vice versa.

For instance, -3 is clearly a root of $g(x) = x^2 - 9$ because $g(-3) = (-3)^2 - 9 = 0$. Therefore, you automatically know that $(x - (-3))$, which equals $(x + 3)$, is one factor of $g(x)$. That's no big surprise for you because $g(x)$ is the difference of perfect squares and factors into $(x + 3)(x - 3)$. When the functions aren't so easy to factor, however, this is a handy relationship to keep in mind.

When you're asked to factor a function or find its roots (according to what I just told you, they amount to the same thing) and its degree is 3 or higher, here's what you should do:

1. **Generate a list of possible roots using the rational root test.** Don't forget to include both the positive and negative versions of each.

2. **Use Descartes' Rule of Signs to narrow the list.** If you find that there can be only one negative root, for instance, once you find that root, you can confidently cross all the other negative numbers off the list of suspects.

3. **Test the roots using synthetic division.** If you get a remainder of 0, the number in question is definitely a root; otherwise, it's not. Once you find a root, use the resulting quotient to find the next root rather than the original function again. (This makes things a little easier because the quotient is always one degree less than the original function, and it's probably shorter as well.)

How'd You Do That?

In step 3, I told you that a remainder of 0 means that the number you're testing is a root of the function. This conclusion is based on a mathematical principle called the **Remainder Theorem,** which makes this claim: If a function $f(x)$ is divided by the binomial $(x - a)$, then $f(a) = r$ (the remainder).

In other words, the remainder obtained when synthetically dividing a function by some value will be the same as the result of plugging that value into the function.

Testing roots in step 3 can get a bit frustrating because you'll usually have to try a bunch until you find one that works. Don't give up—stick with it! If you're totally stuck, try to draw a quick sketch of the function by plotting a bunch of points; remember that a function's roots are also its x-intercepts, so if the points you're plotting are really close to the x-axis, you know you're getting close to a root as well.

Example 5: Calculate all of the roots of the functions:

a. $f(x) = 2x^3 + 3x^2 - 18x + 8$

Solution: Descartes' Rule of Signs tells you that there are either two or zero positive roots, but that there is only one possible negative root. The rational root generates this list of possibilities: $-8, -4, -2, -1, -\frac{1}{2}, \frac{1}{2}, 1, 2, 4, 8$.

You should probably start with the smallest integers first because they're much easier to divide. However, if you try 1 and –1, they both have nonzero remainders. Luckily, 2 does not:

$$\underline{2|}\ \ \begin{array}{rrrr} 2 & 3 & -18 & 8 \\ & 4 & 14 & -8 \\ \hline 2 & 7 & -4 & 0 \end{array}$$

So, 2 is a root; after dividing it out, you get the quotient $2x^2 + 7x - 4$, which can be factored using the Bomb Method (refer to Chapter 4): $(2x - 1)(x + 4)$. To get the final two roots, set each of those factors equal to 0 and solve; you get $\frac{1}{2}$ and –4. Therefore, the roots of $f(x)$ are –4, $\frac{1}{2}$, and 2.

b. $g(m) = m^4 - 3m^3 + 3m - 1$

This time, the possible list of roots is remarkably short; only 1 and –1 could be roots of the function. Descartes' Rule of Signs apparently confirms this, stating that there will be either three or one positive root(s) and only one negative root. Let's see if –1 is that lonely negative root:

$$\underline{-1|}\ \ \begin{array}{rrrrr} 1 & -3 & 0 & 3 & -1 \\ & -1 & 4 & -4 & 1 \\ \hline 1 & -4 & 4 & -1 & 0 \end{array}$$

Excellent, –1 is a root. Now use the bottom row of numbers from that division problem to see if 1 is a root as well:

$$\underline{1|}\ \ \begin{array}{rrrr} 1 & -4 & 4 & -1 \\ & 1 & -3 & 1 \\ \hline 1 & -3 & 1 & 0 \end{array}$$

So, both –1 and 1 are roots, and you're left with the quadratic $m^2 - 3m + 1$, which is not factorable. This means that the remaining two roots must be either irrational or complex. Either way, you can calculate them using the quadratic formula because it works for any quadratic, no matter how ugly its roots:

$$m = \frac{3 \pm \sqrt{9 - 4}}{2} = \frac{3 \pm \sqrt{5}}{2}$$

The roots of $g(m)$ are –1, 1, $\frac{3 + \sqrt{5}}{2}$, and $\frac{3 - \sqrt{5}}{2}$.

You've Got Problems

Problem 3: Calculate all the roots of $g(x) = 3x^3 + 16x^2 + 7x + 10$.

The Least You Need to Know

♦ Quadratic equations can be solved via the factoring method, by completing the square, or through the quadratic formula.

♦ The leading coefficient test describes the end behavior of a function.

♦ Descartes' Rule of Signs helps you estimate how many positive and negative roots a function possesses.

♦ All of the possible rational roots of a function can be listed using the rational root test.

Logarithmic Functions

In This Chapter

♦ Understanding logarithmic notation

♦ Features of a logarithm's graph

♦ Rewriting expressions using logarithmic properties

♦ Applying the change of base formula

There are a bunch of things I'll miss about the 1990s. It was a good decade for me because a lot of my best memories lie nestled in that short 10-year span. I went to college, got a good job, met the woman of my dreams, bought my first house, wrote my first book, and ate my first kiwi (not as momentous, perhaps, as the other events, but good evidence of a sheltered life, at least when it comes to produce). However, the thing I'll miss most about the 1990s is the Taco Bell Chihuahua.

Never before (at least in my memory) had a huge fast food company entrusted its sales to a shaky, pointy-eared, talking Chihuahua, but it worked like gangbusters. She even had her own catchphrase (the dog was a she, but the voice was a male comedian, so gender pronouns are actually a bit complicated, as is sometimes the case when you're writing about talking dogs): "Yo Quiero Taco Bell!" I don't know any Spanish, but I think the rough translation is "Taco Bell gives me really bad heartburn, so I'd recommend going somewhere else to eat dinner."

It's not the phrase I'll miss, nor the dog, nor the heartburn, but the very first commercial, in which the director of the commercial (it may have been Orson Wells) established that the dog could talk by having it walk by a television and answer a question from the television show *Jeopardy!* Believe it or not, the dog's first words, in response to Alex Trebek, were "What is a logarithm?"

A shorter version of the commercial aired more often than the full version, but it didn't feature the math question. I can only assume that's because very few people in America had any idea what the dog was talking about. (I bet that's a phrase you never expected to see in print!) In this chapter, I'll catch you up to the smart, iconic doggie and explain exactly what a logarithm is and why you should care (although I can't guarantee it won't give you heartburn).

Evaluating Logarithms

The *logarithmic expression* $\log_a x$ has *base a* and *argument x;* it's read "log base *a* of *x*," and its value is actually equal to an exponent. Essentially, the equation $\log_a x = y$ can be translated as "When *a* is raised to the *y* power, you will get *x*." Therefore, the logarithmic equation $\log_3 9 = 2$ is true because $3^2 = 9$.

Talk the Talk

The **logarithmic expression** $\log_a x$ has a value that answers the question "To what power must I raise the **base, a,** to get *x* (the **argument**)?" Note that the base of a logarithmic function will be a positive number not equal to 1. (Because 1 to any power just equals 1, a logarithmic base of 1 is wholly uninteresting.)

Even though a logarithmic expression may not make much sense to you at first, these are still pretty easy to evaluate because you can rewrite them in exponential form.

Example 1: Determine the value of *x* in each of the following logarithmic expressions:

a. $\log_{10} 1000 = x$

This equation translates into "10 raised to what power is equal to 1000?" and can be rewritten in exponential form like this: $10^x = 1000$. The answer is $x = 3$.

b. $\log_8 2 = x$

This one is a little trickier. You are essentially solving the equation $8^x = 2$. How in the world can you raise a number to an exponent and end up with something smaller as a result? It's easy if you notice that 8 and 2 are powers of 2, and you rewrite 8 as 2^3 and 2 as 2^1:

$$8^x = 2$$

$$\left(2^3\right)^x = 2^1$$

$$2^{3x} = 2^1$$

Because the two sides are equal, their exponents must be equal, too, so $3x = 1$ and, thus, $x = \dfrac{1}{3}$.

c. $\log_x 32 = 5$

This translates into the exponential expression $x^5 = 32$, and the only number that equals 32 when raised to the fifth power is $x = 2$.

You've Got Problems

Problem 1: Determine the value of x in each logarithmic expression:

a. $\log_5 x = 3$

b. $\log_7 7 = x$

c. $\log_{16} \dfrac{1}{4} = x$

Graphs of Logarithms

Once you know what a generic logarithmic graph, $f(x) = \log_a x$, looks like, you can use the function transformations you learned in Chapter 6 to shift, stretch, and reflect that graph to your heart's content (that is, if your heart can truly be contented by just graphing log functions—I think it's doubtful). In other words, you need to memorize this graph and add it to your repertoire. Before you're done with this book, the number of basic function graphs you'll need to know by heart will have risen from 4 in Chapter 6 to a grand total of 12.

Critical Point

In case you're curious, the 12 basic function graphs you'll learn by the end of the book are: x^2, x^3, $\sqrt{x}$, $\dfrac{1}{x}$, $\log_a x$, a^x, $\cos x$, $\sin x$, $\tan x$, $\cot x$, $\sec x$, and $\csc x$. (If you have no idea what those last four are, don't worry—I'll introduce them in Chapter 11.)

All logarithmic graphs have basically the same shape (and what a great shape it is—they must all work out, I swear), so I'm going to select the base 2 logarithmic function ($f(x) = \log_2 x$), and plot a bunch of points to get a sense of its graph. The first thing you'll notice is that negative arguments for $f(x)$ don't make any sense. For example, $f(-4) = \log_2 (-4)$, which asks the question "2 raised to what power results in a value of -4?" No such value exists; there's no way to turn a positive number into a negative number using only an exponent (even a negative exponent won't affect its sign—it will only generate a reciprocal).

With this in mind, I am going to plug in only positive numbers for x. It will take only a few x values (that I'll carefully hand-select in the chart that follows) to get a good sense of the graph of $f(x)$. Each row starts with the x value plugged in, asks the question "2 to what power gives you that value?", and answers that question in the last column.

x	$f(x) = \log_2 x$	$f(x)$
$\dfrac{1}{16}$	$2^? = \dfrac{1}{16}$	-4
$\dfrac{1}{2}$	$2^? = \dfrac{1}{2}$	-1
1	$2^? = 1$	0
2	$2^? = 2$	1
8	$2^? = 8$	3

Now graph the five coordinate pairs created by the chart: $\left(\dfrac{1}{16}, -4\right)$, $\left(\dfrac{1}{2}, -1\right)$, $(1,0)$, $(2,1)$, and $(8,3)$. You'll end up with Figure 8.1.

Figure 8.1

The graph of f(x) = log₂ x with the five points that I used to sketch the graph highlighted.

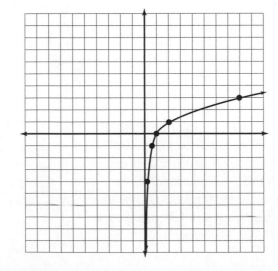

The graph of $\log_2 x$, as well as every other untransformed logarithmic function, has a few important characteristics that not only help you to draw the graph, but also give you a better understanding of logarithms:

- **The graph contains no negative x values.** Remember, there's no exponent that can change the sign of its base. Therefore, negative numbers aren't valid inputs for a logarithmic function, and that's reflected in the graph because it contains no negative x values.

- **The y-axis is a vertical asymptote of the graph** (for more information on asymptotes check out Chapter 6). An argument of 0 might be impossible, but the graph will get infinitely close to that impossible value, represented by the vertical line $x = 0$.

- **The graph is below the x-axis until $x = 1$.** Teensy x values (between 0 and 1) are arguments that are smaller than the base (which is 2), so the output will have to be a negative number (because it represents a negative exponent). For instance, an argument of $x = \dfrac{1}{4}$ translates into the expression $\log_2 \dfrac{1}{4}$, which is equal to -2.

- **The graph contains the point (1,0).** When you plug 1 into a logarithmic function, you're asking "To what power must I raise the base to get a final value of 1?" The only such value is 0 because anything to the 0 power equals 1 (except 0, which is not a valid logarithmic base anyway).

- **The graph grows very slowly once it crosses the x-axis.** That quick and steep growth that occurs between $x = 0$ and $x = 1$ is nowhere to be found once the graph crosses the x-axis. For every single unit higher that the graph of $f(x)$ travels vertically, it must travel a power of 2 horizontally, and exponents grow quickly. Notice in Figure 8.1 that it took only eight horizontal units for $f(x)$ to grow from an infinitely negative height to a positive height of 3 (because $2^3 = 8$). It will then take eight more horizontal units before the graph increases only *one more unit* to a height of 4 (because $2^4 = 16$).

- **The graph reaches a height of 1 at the argument matching its base.** In Figure 8.1, the graph passes through the point (2,1) because the base of the logarithm is 2. If you were graphing $y = \log_{18} x$, the graph would have passed through the point (18,1). Whenever the base and the argument match, you're asking the question "To what exponent must I raise the base so that its value stays exactly the same?" The answer's always 1.

Critical Point

If the base of a log equals the base of its argument, the answer is the exponent of the argument. For instance, $\log_5 5^3 = 3$. If the base and argument are *exactly equal* (as in the expression $\log_4 4$), the expression has a value of 1 (the implied exponent of the input). If you rewrite the logs in exponential form, the reason becomes clear: $\log_5 5^3 = 3$ becomes $5^3 = 5^3$, which is definitely true.

Graphing a logarithmic function requires no new skills, as long as you remember your function transformations.

Example 2: Sketch the graph of the function $g(x) = \log_3 (-x) + 2$.

Solution: Start with the untransformed version of this function: $y = \log_3 x$. All basic logarithms pass through the point (1,0) and have the y-axis as a vertical asymptote. In addition, this function will reach a height of 1 when x equals the base, so the graph of $\log_3 x$ must pass through (3,1). I've drawn the graph of $y = \log_3 x$ as a dotted curve in Figure 8.2. Now it's time to apply the transformations to this graph.

A $-x$ inside the function (instead of x) tells you to reflect the graph across the y-axis; you then must move the entire graph up two units (because of the "+ 2" at the end of the function). Compare the untransformed graph of $y = \log_3 x$ and the final graph of $g(x) = \log_3 (-x) + 2$ in Figure 8.2 to visualize those transformations.

Figure 8.2

The graph of g(x) *is simply the graph of* y = log₃ x *once it's reflected across the* y-axis, *moved up two units, and given a nice-looking haircut.*

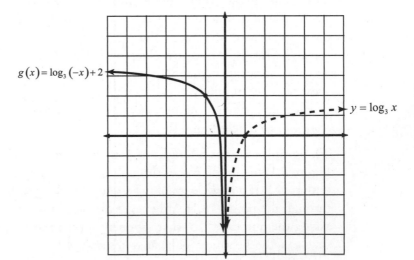

You've Got Problems

Problem 2: Graph the function $f(x) = -\log_5 (x + 3)$.

Common and Natural Logarithms

You'll find all kinds of logs out there: logs from lumber mills, rotted forest logs that are the result of fallen trees, logs with lumberjacks on top trying to spin one another off and into a pond, and even oversized plastic logs you can sit in at an amusement park flume ride, whose cushions are always covered by just enough standing water to soak straight through your shorts and underpants as soon as you sit down in them at the boarding gate.

Math uses even more kinds of logs than that because there are an infinite number of bases you can use in a logarithmic function. If you've ever taken a good, hard look at a calculator, however, you'll notice that it supports only two kinds of logs floating on that infinite sea of possibilities: the *common log* and the *natural log*.

Talk the Talk

The **common logarithm**, written log x and read as "the log of x," is understood to have a base of 10. The **natural logarithm** has an assumed base of e; it is written ln x and can be read "the natural log of x" or "L-N of x."

The common logarithm has base 10. It's used so commonly that a log expression written without a base is assumed to have a base of 10. So, the expressions $\log x^2$ and $\log_{10} x^2$ mean exactly the same thing. Before you get too weirded out about that, remember that you actually omit a lot of things in math and assume that they're understood; for instance, y really means $1y^1$ (even though no exponent or coefficient is stated), and $\sqrt{5x}$ really means $\sqrt[2]{5x}$ (even though the index of 2 was not written in the original square root).

The natural logarithm also has a base that's not written explicitly; it's written ln x so that you don't confuse it with log x. The base of the natural log, however, does not equal 10—it equals e. Do you know what e is? It's a predefined mathematical constant, like i (which equals $\sqrt{-1}$) or π (which equals 3.1415 ...), called Euler's number.

Kelley's Cautions

Euler, whose name is emblazed on the constant e, is pronounced "oil-er," not "you-ler."

Like π, Euler's number is an irrational number (because it never repeats and terminates) that equals approximately 2.71828182845904523536028747135527 .... That's why the notation e is so handy; writing ln 7 sure beats the heck out of writing $\log_{2.71828182845904523536028747135527\,\ldots}7$.

Because these logarithmic bases are encountered the most frequently, you'll notice that scientific and graphing calculators have buttons for them, which makes calculating their exact decimal values a breeze. Calculating those decimal values by hand is as useless an exercise as calculating square roots by hand, so mathematicians—even ones who sneer at technology—gladly punch logarithms into their calculators if decimal values are needed.

Example 3: Simplify the expression $\ln e^2$.

Solution: The base of the logarithm is e, so this expression asks "e must be raised to what power in order to get e^2?" When the base of a log (e) equals the base of its argument (e^2), the answer is the exponent—2, in this case.

Change of Base Formula

If you believed me when I told you that logarithms of base 10 and base e are used the most frequently in precalculus (and you should believe me—I am a relatively honest person), that statement is nonetheless unsatisfying. You'll still run into other bases. It's as if we're on a hike in the woods together, and you're asking me "What happens if I step on a poisonous snake?" and I respond by saying "Don't worry—the most common snakes around here are not poisonous." I might be right, but if you run into one that is, the snakebite venom is not going to course through your veins any slower just because the snake was improbable.

Let me get away from this unnecessarily creepy snake metaphor for a second and explain exactly what I mean. Most of the time, you'll be expected to leave your answers with the logarithms still inside. For instance, a final answer of $\log_7 13$ is completely acceptable because it cannot be simplified any further and cannot be more accurate or clear. However, in some rare instances, you'll need to know exactly what the decimal value of $\log_7 13$ is.

You could use a little logic to guess at its value. The expression $\log_7 13$ asks "7 to what power equals 13?" so you know its value is larger than 1 (because $7^1 = 7$, which is less than 13). You also know that $\log_7 13$ must be smaller than 2 because $7^2 = 49$, which is larger (a whole heck of a lot larger, in fact) than 13. So, using your big, pulsating brain, you could reason that $\log_7 13$ is between 1 and 2 and is probably closer to 1 than 2. Unfortunately, that's still not a very accurate answer.

Luckily, you can use a little conversion trick called the *change of base formula* to calculate logarithms with bases other than 10 and *e*. The formula looks like this:

$$\log_a x = \frac{\log x}{\log a} \qquad \text{or} \qquad \log_a x = \frac{\ln x}{\ln a}$$

In other words, divide the common log of the argument by the common log of the base, or, if you'd rather, divide the natural log of the argument by the natural log of the base. Either way, you'll get the same answer. Therefore, to calculate $\log_7 13$, you divide the common log of 13 by the common log of 7. Remember, your calculator will be more than happy to calculate the log values and do the division for you.

Critical Point

You get the same final value for $\log_7 13$ if you use natural logs instead of common logs:

$$\log_7 13 = \frac{\ln 13}{\ln 7} = \frac{2.56494935746}{1.94591014905} \approx 1.318$$

$$\log_7 13 = \frac{\log 13}{\log 7} = \frac{1.11394335}{0.84509804} \approx 1.318$$

Example 4: Determine the value of $\log_3 5$, rounded to three decimal places.

Solution: Apply the change of base formula using natural or common logs:

$$\log_3 5 = \frac{\log 5}{\log 3} \qquad\qquad\qquad \log_3 5 = \frac{\ln 5}{\ln 3}$$
$$\approx \frac{0.698970004}{0.47712125} \qquad \text{or} \qquad \approx \frac{1.6094379}{1.0986122}$$
$$\approx 1.46497 \qquad\qquad\qquad\qquad \approx 1.46497$$

Don't forget to round your answer to three decimal places, as the problem indicates: 1.465.

You've Got Problems

Problem 3: Determine the value of $\log_2 19$, rounded to three decimal places.

Logarithmic Properties

Logarithmic functions possess three major properties. These properties allow you to manipulate logarithms with two goals in mind: shrink the expressions down into a more compact form (like dehydrating meat into jerky form), or do just the opposite and expand the expressions from compact notation into long, bloated notation. Every property can be used to either condense or bloat, and I'll show you how as I introduce them one by one:

Kelley's Cautions

You can apply log properties only to logarithms with the same base, although it doesn't matter what that base is. Don't get confused. Even though I've written the properties using common logs, any bases can be used, as long as all the bases in the same problem match.

- **log a + log b = log ab.** In other words, the sum of any logs (with the same base) is equal to the log of their product. For instance, log 5 + log 9 = log 45. (If you don't believe me, verify it using a calculator.) You can also use this rule to work backward. For instance, you could rewrite the expression log 50 as "log 25 + log 2" because 25 · 2 = 50. A word of warning: The product of two logs does *not* equal the log of their sum, so (log a)(log b) ≠ log(a + b).

- **log a – log b = log $\frac{a}{b}$.** The difference of two logs (with matching bases) is equal to the log of their quotient. This makes a lot of sense: If the sum of two logs equals one log containing multiplication, then the difference of two logs (the opposite of a sum) equals one log containing division (the opposite of multiplication).

  Using this property, you could rewrite the expression "ln 3 – ln y" as " ln $\frac{3}{y}$ " or go backward and rewrite "$\log_3 2$" as "$\log_3 10 - \log_3 5$" (because 10 ÷ 5 = 2).

- **log a^b = b log a.** You can take the exponent off the quantity you're "logging" and multiply it out in front of the logarithm. Therefore, $\log_3 5^2 = 2\log_3 5$.

How'd You Do That?

Are you wondering how the third log property works, such as why log x^3 = 3log x? You can prove that those two expressions are equivalent very easily. Just think of log x^3 as log ($x · x · x$). According to the first log property, the log of a product can be rewritten as a sum of individual logs:

$$\log (x · x · x) = \log x + \log x + \log x$$

So now you know that log x^3 = log x + log x + log x, but those three terms on the right side of the equals sign are like terms and can be added (log x + log x + log x = 3log x). Therefore, log x^3 = 3log x.

Thanks to these properties, there are lots of ways to write answers to logarithmic problems. For instance, if you got an answer of $x = \log \frac{1}{2}$ to a problem, but the back of your textbook insists that the answer is $x = -\log 2$, don't despair. You actually got the same thing!

Use the final log property to move the book answer's coefficient of -1 back up into the exponent, which changes $x = -\log 2$ into $x = \log 2^{-1}$. A negative exponent tells you to take the reciprocal, which changes $x = \log 2^{-1}$ into $x = \log \frac{1}{2}$, so the answers are equal.

Example 5: Use logarithmic properties to rewrite the expressions:

a. Expand $\ln \frac{xy^2}{z^3}$.

 Solution: The log of a quotient can be rewritten as the difference of the logs of its numerator and denominator:

$$\ln (xy^2) - \ln z^3$$

 The left natural log contains a product (xy^2), which should be expanded into the sum of two logs:

$$\ln x + \ln y^2 - \ln z^3$$

 Now move the exponents of y and z to the front of their respective log expressions via the third log property:

$$\ln x + 2\ln y - 3\ln z$$

b. Compact into a single logarithm: $\frac{1}{3}\log a - 2(\log b + \log c)$.

 Solution: Start by distributing the -2 through the parentheses:

 $\frac{1}{3}\log a - 2\log b - 2\log c$. Now use the last log property to change the coefficients back into exponents, noting that an exponent of $\frac{1}{3}$ is the same as a cube root: $\log \sqrt[3]{a} - \log b^2 - \log c^2$.

 The difference of the first two logs should be rewritten as the quotient of a single log: $\log \frac{\sqrt[3]{a}}{b^2} - \log c^2$. Now divide that fraction by c^2—just toss it into the denominator: $\log \frac{\sqrt[3]{a}}{b^2 c^2}$.

You've Got Problems

Problem 4: Use logarithmic properties to rewrite the expressions:

    a. Expand $\log \dfrac{2}{x^3 \sqrt{y}}$.

    b. Compact into one log: $\ln (x - y) - 3\ln y$.

The Least You Need to Know

    ◆ If $\log_a x = y$, then $a^y = x$.

    ◆ The graph of $f(x) = \log_a x$ passes through the points $(1,0)$ and $(a,1)$; the y-axis is its vertical asymptote.

    ◆ If no base is written for a logarithm, it's understood to be 10; the base of the natural logarithm ($\ln x$) is e.

    ◆ Three logarithmic properties allow you to expand and compact log expressions.

Exponential Functions

In This Chapter

- ◆ Defining exponential functions
- ◆ Investigating the relationship between exponential and logarithmic functions
- ◆ Solving exponential and logarithmic equations
- ◆ Calculating exponential growth and decay

As I grow older, I can feel my life slowly beginning to evolve. I'm doing a lot of the things that I considered horribly uncool when I was younger, like favoring talk radio in the car over music stations. Perhaps I'm wrong, but I think all guys, once they reach a certain age, start to feel the gravitational pull of the curmudgeon—to change into the mean old man that no one likes. Portions of my brain that used to think "Wow, that car is awesome" are morphing and are now thinking things like "Those kids had better get off my lawn or I'm going to throw my shoe at them."

One of the key characteristics of a curmudgeon is the tendency to be impatient about things that really aren't that big of a deal. (This is also a key characteristic of an insane person, but I will try to be a positive thinker here.) One thing, above all else, is beginning to drive me absolutely bananas: Misuse of the word *literally*. How many times have you heard someone say "I was *literally* scared to death"? I hear it all the time, which is weird because if they were *literally* scared to death, it would be much harder to hold a conversation with them, especially at a cocktail party.

Sometimes people will also toss in math terms they don't really understand, to spruce up their tale. When talking to other parents, I might hear, "Little Mandy was short for so long, then once she hit middle school, she *literally* grew exponentially for about three years." That would be impressive, indeed, but unless Mandy is more than 50 feet tall, they can't mean that literally. Sure, exponential growth basically means "fast growth," but there's a whole lot more to it than that. In this chapter, I introduce you to exponential functions, the long-lost cousins of logarithmic functions, and tell you exactly what exponential growth is. And I'll do it all while wearing these brown, knee-high dress socks with my sandals because lately I just can't resist the urge to wear them as an ensemble.

Graphing Exponential Functions

An exponential function is not just a function containing exponents, even though that would make a lot of sense. An *exponential function* is a function with a variable actually inside the exponent itself. Therefore, $f(x) = 5^x$ is an exponential function, whereas $h(x) = x^5$ is not.

Talk the Talk

Exponential functions contain an exponent with a variable inside, such as $g(x) = 2^x$. The base of an exponential function (in this case, 2) is a positive number other than 1, just as with logarithmic function bases. We'll focus on exponential functions whose bases are integers larger than 1.

To get a good idea of what an exponential graph looks like, I'll sketch the graph of a basic one: $g(x) = 2^x$. The easiest way to do this is to plug in a bunch of numbers for x, which I've done in the following chart:

x	$g(x) = 2^x$	$g(x)$
-4	$2^{-4} = \dfrac{1}{2^4} = \dfrac{1}{16}$	$\dfrac{1}{16}$
-1	$2^{-1} = \dfrac{1}{2}$	$\dfrac{1}{2}$
0	$2^0 = 1$	1
1	$2^1 = 2$	2
3	$2^3 = 8$	8

Based on a few unique x values and some simple arithmetic, you now know that the points $\left(-4, \dfrac{1}{16}\right)$, $\left(-1, \dfrac{1}{2}\right)$, $(0,1)$, $(1,2)$, and $(3,8)$ belong on the graph of $g(x)$. Plot those points, as I've done in Figure 9.1, to get a nice sketch of the graph.

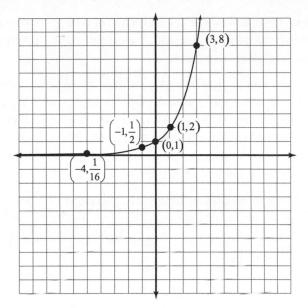

Figure 9.1

The graph of g(x) = 2^x, a typical exponential function.

Like logarithmic function graphs, all exponential function graphs resemble one another closely. No matter what the base of the exponential function is, its graph has these characteristics:

♦ **Negative *x*-values result in teeny, tiny numbers that lie close to the *x*-axis.** A negative exponent translates into a reciprocal, so a very negative exponent means a very small fraction. For instance, if $g(x) = 2^x$, then $g(-100) = 2^{-100} = \dfrac{1}{2^{100}}$, which equals approximately .0000000000000000000000000000078, a teeny, tiny number indeed.

♦ **The *x*-axis acts as a horizontal asymptote.** No matter how enormously negative the number you plug into an exponential function is or how small the output is, you'll never actually get 0. Because an output of 0 is impossible, the graph will never cross the *x*-axis.

♦ **The graph contains the point (0,1).** Since any base raised to the 0 power equals 1, plugging 0 into an exponential function always gives an output of 1.

Critical Point _____

If an exponential function has a base of e (Euler's number), it's called the **natural exponential function,** just like a logarithm with base e is called the natural logarithmic function.

◆ **The graph never goes below the *x*-axis.** The outputs (*y*-values) of an exponential graph are always positive because you cannot raise a positive number to any exponent and get a negative result.

◆ **The graph will rise steeply and quickly once it passes the *y*-axis.** Each unit you move right along the *x*-axis translates into an exponential vertical increase on the *y*-axis. Look at Figure 9.1 again. When $x = 4$, the graph will be $2^4 = 16$ units high—that's twice as high as the graph was able to reach on the *entire x-interval* $(-\infty, 3]$. It doubled its height in the space of just one horizontal unit! When $x = 5$, the function is 32 units high, and it just grows faster and faster after that.

◆ **When *x* = 1, the graph reaches a height that matches its base.** Plugging $x = 1$ into the function really means raising the base to the 1 power, which won't change its value. Therefore, the graph of the exponential function $f(x) = a^x$ always contains the point $(1, a)$.

Use these characteristics and your knowledge of graphical transformations to quickly sketch the graph of simple exponential functions.

Example 1: Sketch the graph of $f(x) = -4^{x+3} - 2$.

Solution: Start by graphing the exponential function $y = 4^x$. According to the preceding list, the graph will have a horizontal asymptote at the *x*-axis, will pass through $(0,1)$, and then will rise very steeply through the point $(1,4)$. I've drawn the graph of $y = 4^x$ in Figure 9.2 as a dotted curve.

Three transformations are applied to the graph of $f(x)$: a reflection across the *x*-axis (thanks to the negative sign in front of 4^{x+3}), a horizontal shift three units to the left (thanks to the 3 added within the exponent), and a horizontal shift two units down (thanks to the 2 subtracted from 4^x). The final graph appears as a solid curve in Figure 9.2.

You've Got Problems
Problem 1: Sketch the graph of $h(x) = 3^{-x} + 1$.

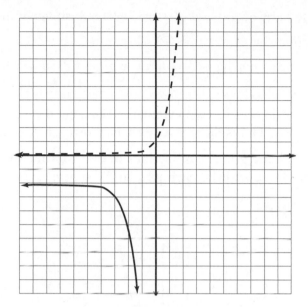

Figure 9.2

The untransformed graph of $y = 4^x$ (the dotted curve) becomes the graph of $f(x) = -4^{x+3} - 2$ once it is horizontally reflected, shifted 3 units left, and then moved 2 units down.

The Logarithmic/Exponential Balance of Power

Have you noticed the eerie similarities between exponential and logarithmic functions? Isn't it kind of strange that in log functions, your output is an exponent, but in exponential functions, the exponent is the input? Even weirder are their asymptotes—logarithmic functions have a y-axis asymptote, and exponential functions have an asymptote along the x-axis. It's almost as if these functions were mirror images of each other in some bizarre way, and it warrants a little deeper investigation.

The unnerving relationship between log and exponential functions that's nibbling away at your subconscious exists because they actually are inverse functions. More specifically, logarithms and exponential functions with the same base (such as $\log_2 x$ and 2^x, for example) are inverses of one another. I picked base 2 there because you actually know a thing or two about the functions $f(x) = \log_2 x$ and $g(x) = 2^x$ already; I graphed $f(x)$ in Chapter 8 (see Figure 8.1) and $g(x)$ just a few pages ago (see Figure 9.1) by plotting a bunch of points for each.

Allow me to present three major pieces of evidence (based on the characteristics of inverse functions I listed in the section "Inverse Function Behavior" in Chapter 6), just in case you're not convinced that these functions are, indeed, inverses:

- **The coordinate pairs of f(x) and g(x) are reversed copies of one another.** These are the points I used to graph $f(x) = \log_2 x$ in Chapter 8: $\left(\frac{1}{16}, -4\right)$, $\left(\frac{1}{2}, -1\right)$, (1,0), (2,1), and (8,3). Flipping back only a few pages will show you that the points I used to graph Figure 9.1 ($g(x) = 2^x$) are $\left(-4, \frac{1}{16}\right)$, $\left(-1, \frac{1}{2}\right)$, (0,1), (1,2), and (3,8). The numbers in each pair are exactly the same, but the x and y in each have swapped places. This is the signature characteristic of an inverse function.

- **The graphs of f(x) and g(x) are reflections of one another across the line y = x.** For convenience, I drew both graphs on the same coordinate plane in Figure 9.3.

Figure 9.3

All inverse functions must be reflections of one another across the line y = x, *just as the graphs of* f(x) = log₂ x *and* g(x) = 2ˣ *are.*

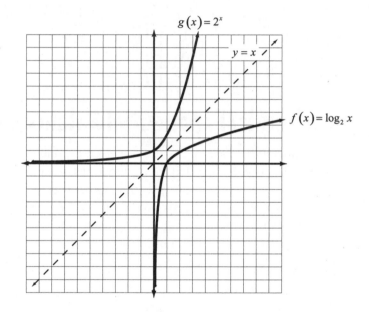

$g(x) = 2^x$

$y = x$

$f(x) = \log_2 x$

Kelley's Cautions

Even though it's important for f(x) and g(x) to have reversed ordered pairs and to be reflections of each other across the line y = x, the mandatory requirement for inverse functions is that they make the equation f(g(x)) = g(f(x)) = x true. As long as they compose this way with one another, the coordinate pair and graph requirements will follow automatically.

♦ **When $f(x)$ and $g(x)$ are composed with one another, everything cancels out except x.** In other words, these functions fulfill the most essential requirement of inverse functions: $f(g(x)) = g(f(x)) = x$. The expression $f(g(x)) = \log_2 (2^x)$ asks the question, "To what power should you raise 2 so that 2^x is the result?"; the answer is x. You could also use the shortcut I described in Chapter 8: When the bases of the log and its argument are equal, the expression equals the argument's exponent, which is x, in this case.

It's a little harder to prove that $g\big(f(x)\big) = 2^{\log_2 x}$ also equals x. If you're curious about it, check out the nearby "How'd You Do That?" sidebar. If you're not too curious (and fear getting confused), just trust me—it does equal x.

These justifications that ensure $f(x) = \log_2 x$ and $g(x) = 2^x$ are inverses hold, even if the base isn't 2. Just as long as the bases (a) are equal, $f(x) = \log_a x$ and $g(x) = a^x$ are inverse functions.

How'd You Do That?

To prove that $g\big(f(x)\big) = 2^{\log_2 x}$ actually just equals x, you should first set the exponent equal to a neutral variable. I'll use y: $\log_2 x = y$. That expression can be rewritten in exponential form: $2^y = x$.

Here's the cool part: Substitute the expression $\log_2 x$ into that equation for y (which is allowed since you explicitly stated that $\log_2 x = y$), and watch what happens:

$$2^y = x$$
$$2^{\log_2 x} = x$$

This proves that $2^{\log_2 x} = x$, so $g(f(x)) = x$. You've cleared the final hurdle and proved that $f(x)$ and $g(x)$ are inverse functions.

Solving Logarithmic and Exponential Equations

You might be thinking, "Okay, great, logs and exponents with the same base are actually inverse functions, but who cares? What practical use does that information have?" I'm glad you (pretended that you) asked! Here's the answer: It allows you to solve equations containing exponential or logarithmic expressions.

You read that right. I'm talking about equations here, not roots of functions like you did in Chapter 7. You're going to briefly revisit that happy algebraic land that you knew well and loved, the land where as long as you did the same thing to both sides of an equation, just about anything was allowed.

This time, instead of just adding or subtracting the same thing from both sides to move things around, or dividing both sides by the same thing to eliminate coefficients, you'll be applying logarithmic and exponential functions to both sides of an equation. Now that you know that they are inverse functions, you'll exploit that relationship by making them cancel out one another. Let me show you what I mean through a few examples.

> **Critical Point**
>
> Remember, $\ln x$ and e^x have the same base (e), so when they're composed with one another, they'll cancel each other out, leaving only x, just like $\log_a x$ and a^x do. In other words, $\ln e^x = e^{\ln x} = x$.

Example 2: Solve the following equations:

   a. $3^x = 12$

> **Solution:** This problem is tricky because the x you're solving for is trapped as an exponent. To set it free, you'll have to eliminate its base of 3. Since 3^x is an exponential function, you can do just that with its inverse function, a logarithm with the same base. So, take the $\log_3$ of both sides of the equation:

$$\log_3 3^x = \log_3 12$$

> The left side of the equation simplifies to x (since you've got inverse functions plugged into one another).

$$x = \log_3 12$$

> This is an acceptable final answer, but if you'd like to express it as a decimal, you should apply the change of base formula:

$$x = \frac{\ln 12}{\ln 3} \approx \frac{2.484907}{1.098612} \approx 2.262$$

> **Kelley's Cautions**
>
> Try not to round decimals in a problem until you reach the answer. The earlier you round and the fewer decimal places you use when you do, the more inaccurate your answer will be.

b. $\log x - \log 2 = 9$

Solution: To solve for x, you have to eliminate the log function that's attached. Use the logarithmic properties from the end of Chapter 8 to rewrite the left side of the equation as a single log:

$$\log \frac{x}{2} = 9$$

Since this is the common log, its base is 10 and its inverse function must be 10^x. In a process called *exponentiating*, you make each side of the equation an exponent of 10:

$$10^{\log \frac{x}{2}} = 10^9$$

I know that looks very strange, but the only way to cancel out a logarithm is to make it a power of an exponential function with a matching base, and exponentiation is the way to do it.

Now that you've got a common log plugged into an exponential function with base 10, they cancel out one another, leaving only what's inside:

$$\frac{x}{2} = 10^9$$

> **Talk the Talk**
>
> The process of changing the sides of an equation into exponents of a matching base is called **exponentiation**. For instance, exponentiation changes the equation $\log_5 (3x) = 4$ into $5^{\log_5(3x)} = 5^4$. Notice that the exponential base that's introduced is equal to the base of the logarithm so that the inverse functions on the left side cancel, resulting in $3x = 5^4$.

Multiply both sides by 2 to solve for x:

$$x = 2(10^9) = 2,000,000,000$$

c. $6e^{3x} - 4 = 11$

Isolate the exponential term by adding 4 to each side of the equation:

$$6e^{3x} = 15$$

Divide by 6 on both sides to eliminate the coefficient:

$$e^{3x} = \frac{5}{2}$$

Now that you've got the exponential term all by itself and without a coefficient, you can eliminate it by taking the natural log of both sides of the equation:

$$\ln e^{3x} = \ln \frac{5}{2}$$

$$3x = \ln \frac{5}{2}$$

Divide both sides by 3 to finish:

$$x = \frac{\ln \frac{5}{2}}{3}$$

You may not like the complex fraction in there (the fraction is, itself, in a fraction), but there's nothing wrong with it. If you really wanted to, you could use log properties to rewrite the numerator as a difference of natural logs, or you could get a slightly less accurate answer by typing the expression into a calculator to get a decimal equivalent:

$$x = \frac{\ln 5 - \ln 2}{3} \quad \text{or} \quad x \approx \frac{.9162907319}{3} \approx 0.305$$

d. $\ln x + \ln (x - 2) = \ln 35$

Rewrite the left side of the equation as a single natural logarithm using log properties:

$$\ln x(x - 2) = \ln 35$$

$$\ln (x^2 - 2x) = \ln 35$$

Exponentiate both sides of the equation using e, since that's the base of the natural logarithm:

$$e^{\ln\left(x^2 - 2x\right)} = e^{\ln 35}$$

$$x^2 - 2x = 35$$

You're left with a simple quadratic that can be solved via the factoring method:

$$x^2 - 2x - 35 = 0$$
$$(x - 7)(x + 5) = 0$$
$$x = 7 \text{ or } x = -5$$

Uh oh, hold on a minute. If you plug $x = -5$ back into the original equation to check it, you get this:

$$\ln (-5) + \ln (-7) = \ln 35$$

![CAUTION] **Kelley's Cautions**

The final step when solving any logarithmic equation should be to make sure the solutions are valid. Remember, logs can't have negative arguments.

Unfortunately, the arguments of logarithmic functions aren't allowed to be negative, so that solution must be discarded. The only answer to this problem is $x = 7$.

You've Got Problems

Problem 2: Solve the equations:

a. $\log_2 x - \log_2 3 = 8$

b. $8^{x-1} = 5$

Exponential Growth and Decay

In the real world, not many things can truly experience exponential growth, at least for a prolonged period of time. The reason is simple: Exponential growth usually results in really big, unrealistic numbers that stretch the bounds of reality. For instance, if the grizzly bear population in a given area rises exponentially, grizzly bears will eventually overrun the town. Whereas before you might have seen only one or two bears a month, suddenly they're pouring through the town like a furry, brown river. Even the nice old man who ran the ice cream shop has gone missing, and the grizzly bear that's now the proprietor there has a look in his big bear eyes that says "Don't even ask what happened to Mr. Mendelson. You don't want to know."

Eventually, the bears will run out of groceries, crops, and ice cream store owners to eat. A small town cannot support half a million hungry bears who can't understand the intricacies of a market economy. Even worse, the sanitary conditions of the town will undoubtedly degenerate quickly, thanks to what bears used to do only in the woods, according to the old cliché.

To avoid these unrealistic situations, math teachers and textbooks tend to stick to three major kinds of exponential growth and decay problems:

◆ Continuously compounded interest (like you'd receive on a bank checking account, which, unfortunately, will *not* grow at an out-of-control rate)

◆ Bacteria growth (bacteria colonies are so small that thousands upon thousands can grow on a single Petri dish without being limited by the available resources)

◆ *Half-life* problems (exponential *decay* involves quantities getting smaller, not larger)

Talk the Talk _____

The **half-life** of a radioactive element is the length of time it takes for the mass of that element to halve. For instance, if you buy 300 grams of an element whose half-life is 2 years, you'll have only 150 grams of it left exactly 2 years after the purchase date. Two years later, you'll have 75 grams.

The formula for all exponential growth and decay problems looks like this:

$$F = Ne^{kt}$$

Here's what all those variables represent: the quantity you start with (N), the quantity you end up with (F), the rate at which the quantity changes (k), and the amount of time that passed (t). The e in the formula isn't a variable. It's your old friend Euler's number, so exponential growth is actually just a natural exponential function.

Kelley's Cautions _____

Except in interest problems, you won't know what k is right away. In fact, your first job will be to calculate k based on the given information, like in Example 4. In interest problems, though, k will be the interest rate expressed in decimal form. For instance, if an account has a 2.5 percent interest rate, then $k = .025$. Move the decimal two places to the left and drop the percent sign to get k.

Example 3: *(From a bank brochure)* Thanks for opening up a new savings account at Ebenezer Bank and Trust, where our motto is "You'll get no more coal today, Cratchit!" You have chosen our third account type, "Saving for Christmas Future," which compounds continuously at a rate of 0.9 percent and comes with a free creepy tombstone with your name on it. You may think the interest rate tiny, Tim, but an initial investment of $4,000 left untouched for 20 years will yield a return of $ ... (The rest of the brochure is illegible. What number should appear at the end of that sentence?)

Solution: You should apply the formula $F = Ne^{kt}$ because the interest is continuously compounded. Set $N = 4,000$ (it's the original investment, also called the "principal"), $k = .009$ (the interest rate expressed as a decimal), and $t = 20$ (the number of years the money was invested). You don't know what F is. That's what the problem asks you to find:

$$F = 4,000e^{.009(20)}$$

$$F = 4,000e^{0.18}$$

Type this right into your calculator to get an answer. You should round your final answer to two decimal places, since you're dealing with money: F = $4,788.87. You earn less than $800 in interest in 20 years. Better invest somewhere else.

Example 4: Once she got out of her macramé and crocheting phases, your mom fell into an arguably stranger hobby: collecting radioactive isotopes. She's never been happier, but she's also never been watched quite so closely by the FBI. If you buy her 1,000 grams of the isotope she *really* wants, how much will be left on her birthday, which is 30 days from now, if the half-life of the isotope is 8 days?

Kelley's Cautions

Keep your units of time consistent. If you calculate k by measuring t in hours, continue to measure t in hours for the rest of the problem.

Solution: This is not an interest problem, so k is not a simple number plucked out of thin air. (Even though you may be tempted to think so, $k \neq \frac{1}{2}$.) To determine what k is, pretend that one half-life period (in this case, 8 days) has passed. If you started with 1,000 grams, you'll have 500 (exactly half of that) in 8 days, so $N = 1,000$, $F = 500$, and $t = 8$.

Critical Point

If k from the formula $F = Ne^{kt}$ is negative, the problem involves exponential *decay*. However, if k is positive, you're calculating exponential *growth*.

$$F = Ne^{kt}$$
$$500 = 1000e^{k(8)}$$

Solve for k. Start by dividing both sides of the equation by 1,000.

$$\frac{1}{2} = e^{8k}$$

Ah ha! There's a $\frac{1}{2}$ in the problem after all, even though it's not k.

$$\ln \frac{1}{2} = \ln e^{8k}$$

$$\ln \frac{1}{2} = 8k$$

$$k = \frac{\ln \frac{1}{2}}{8}$$

$$k \approx -0.086643397569993$$

Notice that I included lots of k's decimal places so that my final answer is as accurate as possible. At this point, you know that Mom's dangerous element will decay according to this equation: $F = 1000e^{-0.086643397569993(t)}$. To figure out how much is left 30 days after the purchase date, plug in 30 for t:

$$F = 1000e^{-0.086643397569993(30)}$$

$$F = 1000e^{-2.59930192709979491}$$

$$F \approx 74.325 \text{ grams}$$

You've Got Problems

Problem 3: A scientist notes that there are approximately nine bacterial colonies on a Petri dish at 8:00 A.M. Monday morning; by 4:00 P.M. on Tuesday, that number has grown to 113 colonies. Assuming that the growth is exponential, how many colonies will be present Friday at 5 P.M., when the scientist goes home for the weekend to party like only scientists can? (Hint: Measure t in hours.)

The Least You Need to Know

♦ Exponential functions have variables inside their exponents.

♦ Exponential and logarithmic functions with the same base are inverse functions of one another.

♦ The expressions $\log_a a^x$, $a^{\log_a x}$, $\ln e^x$, and $e^{\ln x}$ all equal x.

♦ The formula for exponential growth and decay is $F = Ne^{kt}$. If k is negative, the problem involves exponential *decay*; however, if k is positive, you're calculating exponential *growth*.

Part 3

Trigonometry

In this part, you'll briefly revisit the happy land of geometry and stroll among its curved and angular inhabitants once again. Specifically, you're going to focus on one geometric shape: the triangle. In fact, the measurement of triangles is so important that it has its own special name: trigonometry. Even though you've dealt with triangles before, trigonometry has a taste entirely its own, a spicy mix with distinct flavors you'll be able to pick out, including familiar flavors (like equations and right triangles) and surprisingly zingy new ones (like the unit circle and trigonometric identities). For the next five chapters, you'll enjoy a buffet of trigonometry (or "trig," for short) intended to keep your hunger satiated not only for your precalculus class, but also for any calculus course that may or may not be looming in your future.

Chapter 10

Tangling with Angles

In This Chapter

- ◆ Graphing angles in standard position
- ◆ Measuring angles in radians and degrees
- ◆ Generating coterminal angles
- ◆ Introducing sine, cosine, and tangent
- ◆ Calculating trigonometric values with and without a calculator

When I decided to become a high school math teacher, I knew that lots of people disliked math; it's not as if that was breaking news or anything. Most people look at math as a necessary evil, like vaccinations—painful but supposedly helpful. Unfortunately, disease inoculations have one benefit that math doesn't: You know why they're useful. No one ever asks, "When will I ever need this rubella vaccination in real life?" because the answer's simple: in case you run into someone who has rubella.

Of course, when my high school students asked me, "Why do I have to learn the angles in the unit circle?" I could have told them, "Because otherwise, you'll probably die," but I doubt they would have believed me. Once you get to precalculus, it gets harder to convince students that what they're learning is worthwhile. As a result, when asked by students why they have to learn a given precalculus topic, most teachers rely on one of three arguments:

(1) because it's good for you; (2) because I said so; or (3) if you don't stop asking me that, I am going to triple your homework and throw my 17,000-page Teacher's Edition textbook at your head.

What I didn't know was that many of my fellow (nonmath) teachers were actually pretty good at math when they were students. Most of them were great at elementary math and actually enjoyed algebra, but they started to really resent math when they got to geometry and trigonometry. It's probably because both of those subjects require a slightly different math mindset. There are some bizarre concepts (like proofs) and some definitions that you won't understand the first time you read them (like radian and cosine). To succeed, you can't just memorize the steps for the problem, as you could with most of algebra. And if no one explains these weird concepts to you, you'll get lost in no time.

In this chapter, I give you a good foundation upon which you'll build a solid knowledge of trigonometry. I'll make sure you understand what's happening as we go along so you don't turn into a statistic and end up loathing math just because it started to look a little different. Sure, things may feel a little weird at first, but that's normal because the closer we get to calculus, the stranger (and more interesting) the topics become.

Angles in the Coordinate Plane

In geometry, you learned that an *angle* is made up of two rays (called the *initial side* and the *terminal side*) that share the same endpoint (called the *vertex*), like two straight roads branching off from the same intersection but stretching infinitely long from there. Though you did your fair share of calculations in geometry, the actual measurements of angles often didn't matter. You could prove triangles congruent without actually popping out a protractor to measure their pieces.

In trigonometry, you'll spend a lot of time calculating angle measurements, so most of the time, angles are drawn on the coordinate plane (which makes measuring much easier). For consistency, you should draw angles in *standard position*, meaning …

1. The vertex lies on the origin.

2. The initial side lies on the positive (right) part of the *x*-axis.

3. A positive angle travels counterclockwise from initial to terminal side; a negative angle travels clockwise.

Talk the Talk

An **angle** is created by joining two rays (called the **initial side** and the **terminal side** of the angle, although it doesn't really matter which is which) together at their endpoints, creating a common endpoint called the **vertex**. An angle in **standard position** is graphed on the coordinate plane so that the vertex lies on the origin and the initial side overlaps the positive *x*-axis. If the terminal side also lies on an axis (as with a 90°, 180°, or 270° angle), it is described as **quadrantal**.

Figure 10.1 shows angle $\angle ABC$ graphed in standard position. Notice that the vertex, B, is located at (0,0), and one of its sides, $\left(\overrightarrow{BA}\right)$, lies on the positive *x*-axis. What is the approximate measure of $\angle ABC$ in degrees? It depends upon which way you go (either clockwise or counterclockwise) from the initial side—you could say either $m\angle ABC \approx 315°$ or $m\angle ABC \approx -45°$; both are good approximations.

You might be thinking, "Wait a minute! I thought angles could have only one measurement! How can an angle be both positive and negative?" Technically, you're right— an angle can't be both. However, you need to think outside the box for a minute. We're not actually interested in the angle for now, just its terminal side, and lots of angles will have the same terminal side as $\angle ABC$. More on this comes later in the chapter, when I discuss things called coterminal angles.

How'd You Do That?

Did you ever wonder why the quadrants of the coordinate plane are numbered so strangely? The pattern follows the path of an angle in standard position, beginning in the upper-right quadrant of the coordinate plane and rotating counterclockwise from there.

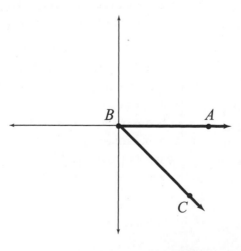

Figure 10.1

$\angle ABC$ *is drawn in standard position because its vertex is on the origin and one of its sides overlaps the positive* x-*axis.*

Measuring Angles

There are lots of units to help you describe an angle's measure, but the one you're probably most familiar with is the *degree*. A degree is simply ¹⁄₃₆₀ of a circle, and despite that weird and somehow disappointingly simple definition, degrees are an undeniable part of American dialect. Slam dunks in basketball requiring the player to spin around once in the air are called "360s"; businesspeople who want to completely change the way they approach things are said to "do a 180" (head in the opposite direction). My friend Rob was driving on an icy and treacherous Pennsylvania road one winter and managed to spin his car completely around, narrowly missing a head-on collision, an incident he references in his occasional declaration, "You haven't lived until you pull a 360 in front of a dairy truck and survive to tell the tale."

Talk the Talk _____

One **degree,** a unit of angle measure, is equivalent to ¹⁄₃₆₀ of a circle and is denoted with a small circle, like this: 360°. Angles less than 90° are described as **acute.** Angles larger than 90° but less than 180° are called **obtuse.** If an angle measures exactly 90°, it's called a **right angle;** angles measuring 180° are **straight angles.**

Degrees and Radians

As comfortable as you may be with degrees, I need to rattle your world a little bit. (You may want to sit down for the bad news.) Modern calculus students usually measure angles using units called *radians,* and since precalculus is the calculus on-ramp, that means you'll have to do the same thing. The good news is that angles measured in radians aren't written as long and unwieldy decimals (like degrees often are), but the bad news is that's because radians are usually written as fractions.

Critical Point _____

In geometry, you named most angles using either one or three letters (like $\angle E$ or $\angle DEF$), but in precalculus, Greek letters are usually used to represent angles. The most common are θ (theta, pronounced "THAY-tuh"), α (alpha, pronounced "AL-fuh"), and β (beta, pronounced "BAY-tuh").

To make things even worse (just for a moment, I promise, and then the clouds will sweep away and things will be bright and cheery again), radians aren't as intuitive as degrees. You'll never hear a skateboarder say, "I just pulled some wicked air and nailed a 4π radians spin," when a "720" is easier to interpret as two rotations. However, it's important to know what the heck a radian is.

Take a look at the circle in Figure 10.2. It's centered at the origin, and I've high-lighted a radius of the circle, the segment connecting the origin to the circle in the first quadrant. I could have drawn the radius anywhere on the circle, but I chose there because if you think of that radius as the terminal side of an angle in standard position (the angle θ in Figure 10.2), it has the same length as the arc of the circle it intersects. In other words, the radius (even though it's straight) is exactly as long as that dark arc (even though it's curved), and that means angle θ measures exactly 1 radian.

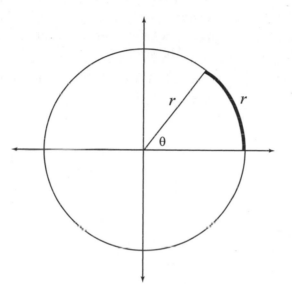

Figure 10.2

When angle θ intersects the circle, it carves out the dark-ened arc, whose length is exactly equal to the radius of the circle. Because of this, θ measures exactly 1 radian. See the connection—rad<u>ius</u> and rad<u>ian</u>?

As you can probably tell just by looking at Figure 10.2, radians are much bigger than degrees. In fact, 1 radian measures about 57.296°, so there are just over 6 radians in a circle, as opposed to 360°. In fact (and this is the cool part), there are exactly 2π radians in a circle. Therefore, 360° and 2π radians are two ways of measuring the same thing, like 0°C and 32°F both measure the same thing (the freezing point of water, and the exact temperature at which you realize your mom was right all those years—you should have worn a hat outside because your ears are so cold, they feel like they're going to break off).

Talk the Talk

An angle that cuts out (or "subtends") an arc of a circle whose length is equal to the radius of that circle (as in Figure 10.2) measures exactly 1 radian, approximately 57.296°.

Converting Between Degrees and Radians

Until you get used to radians, you'll probably need to convert back and forth between them and degrees pretty often. Fortunately, the conversion formula is really easy—all you do is multiply the angle measurement by either $\frac{\pi}{180}$ (to convert from degrees to radians) or $\frac{180}{\pi}$ (to convert from radians to degrees). (Since π and $180°$ both equal the same angle, it's the same thing as multiplying by 1, the multiplicative identity.) As the book goes on, you'll see fewer angles in degrees and more in radians, so force yourself to make the switch.

Critical Point

Here's one way to remember when to multiply by $\frac{\pi}{180}$ and when you need to use $\frac{180}{\pi}$:

The units you want to convert *to* should be on the *top* of the fraction. For instance, in Example 1(a), you're trying to convert to radians, so π should be in the numerator, not 180.

Kelley's Cautions

Whereas degrees have their own denotation (°), radians don't. You could just write "radians" next to your answer, but if the angle measurement contains a π and has no degree symbol next to it, it's safe to assume that the angle is radians.

Example 1: Convert the angle measurements as indicated:

a. Express 135° as radians.

Solution: Multiply 135 by $\frac{\pi}{180}$, and leave the result in fraction form:

$$\frac{135}{1} \cdot \frac{\pi}{180} = \frac{135\pi}{180}$$

You can simplify this fraction (divide the numerator and denominator by 45). You'll get a final answer of $\frac{3\pi}{4}$.

b. Express $-\frac{7\pi}{8}$ as degrees.

This time, you multiply by $\frac{180}{\pi}$. Don't do anything else differently, even though the angle is negative.

$$-\frac{7\pi}{8} \cdot \frac{180}{\pi} = -\frac{1260\pi}{8\pi} = -\frac{1260}{8}$$

Simplify the fraction or rewrite as a decimal to get a final answer of $-\frac{315}{2}°$ or $-157.5°$.

You've Got Problems

Problem 1: Convert the angle measurements as indicated:

    a. Express −150° as radians.

    b. Express $\frac{4\pi}{3}$ radians as degrees.

    c. Express 5 radians as degrees.

Coterminal Angles

The most important feature of an angle drawn in standard position is its terminal side. After all, every angle in the world has the same vertex and initial side if it's in standard position, so those are not very interesting. That terminal side is so important, in fact, that any two angles that have the same terminal side are given a special classification; they're known as *coterminal angles*.

For just a minute, I'm going to go back to an example I introduced near the beginning of this chapter, an angle measuring 315° drawn in standard position, whose terminal side is featured in Figure 10.3.

Talk the Talk

Coterminal angles have the same terminal side when drawn in standard position.

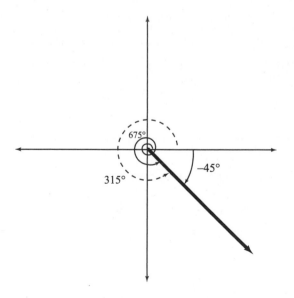

Figure 10.3

The angles 315°, 675°, and −45°, when drawn in standard position, have the same terminal side, so they are coterminal angles.

That terminal side is highlighted for a reason. I already told you that you could go 45° in the negative (clockwise) direction, and you'll end up with the same ray, but those aren't the only two angles that share that terminal ray. What if you went *all the way around the origin* and *then* went 315°? The angle would measure 360° + 315° = 675° (one full circle around, plus the additional 315°).

Think about the game show "Wheel of Fortune." An average person could probably spin the wheel one full rotation, 360° or 2π radians. Maybe Superman could spin the wheel so that it completed 500 rotations (500 · 360° = 180,000° or 2π · 500 = 1000π radians). Either way, both players (super and nonsuper) would end up on the same spot, just as the coterminal angles 315°, –45°, and 675° did.

Therefore, the only thing you have to do to calculate coterminal angles is to add multiples of 360° or 2π radians. The size of the angles (and the length of time it takes for Superman's ridiculously long spin to finally come to an end) may change, but the final result remains unchanged.

Example 2: Calculate two positive and two negative coterminal angles for each of the following:

a. 210°

> **Solution:** Add 360° to get a coterminal angle of 570°, and add 360° more for another coterminal angle of 930°. Subtract 360° from the original angle of 210°, and the result's a negative coterminal angle: 210° – 360° = –150°. Subtract 360° more for another negative coterminal angle of –510°.

Critical Point

Some textbooks like you to write *all* possible coterminal angles for a given angle, and even though there are an infinite number, you can do it. Every angle that's coterminal to $\frac{\pi}{2}$ radians comes from the formula $\frac{\pi}{2} + 2k\pi$, where k is an integer. This is just the fancy way of saying $\frac{\pi}{2}$ plus or minus every multiple of 2π gives you a coterminal angle. This makes sense because adding and subtracting 2π over and over is exactly how we calculated coterminal angles.

To calculate all possible coterminal angles for an angle measured in degrees, use the same reasoning, but change 2π to 360°. For instance, the coterminal angles to 45° are 45° + 360k, where k is once again an integer.

b. $\dfrac{13\pi}{6}$ radians

Solution: Add 2π twice to get your positive coterminal angles. You'll need to use common denominators, so although you're adding 2π, it'll look like you're adding $\dfrac{12\pi}{6}$:

$$\dfrac{13\pi}{6} + \dfrac{12\pi}{6} = \dfrac{25\pi}{6} \quad \text{and} \quad \dfrac{25\pi}{6} + \dfrac{12\pi}{6} = \dfrac{37\pi}{6}$$

To get negative coterminal angles, you actually have to subtract 2π from $\dfrac{13\pi}{6}$ three times. (The first time you do it, you'll get another positive angle, $\dfrac{\pi}{6}$, which is actually a third positive coterminal angle.) Subtracting 2π two more times gives you $-\dfrac{11\pi}{6}$ and $\dfrac{-23\pi}{6}$.

You've Got Problems

Problem 2: Calculate one positive and one negative coterminal angle for each:

a. –900°

b. $\dfrac{3\pi}{4}$ radians

Right Triangle Trigonometry

Let's talk about right triangles for a moment. Of course, when I say "right triangles," I mean triangles that contain a right angle, not triangles that always vote conservatively at the Republican National Convention (those are right-wing triangles, which is an entirely different matter). Did you know that the angles in a right triangle are related to the lengths of its sides?

Three functions govern the relationship of the pieces of a right triangle: sine, cosine, and tangent (usually abbreviated sin, cos, and tan in problems). They work like this: You input an angle into one of the three functions,

Critical Point

Students often use the word "SOH-CAH-TOA" (pronounced "SO-kah-TOE-ah") to remember which sides go with which ratio. (The letters are short for "sine equals opposite over hypotenuse, cosine equals adjacent over hypotenuse, and tangent equals opposite over adjacent.") I prefer the more bizarre mnemonic phrase of my own design: "Cows and hens sleep on hay throughout our apartment."

and the output is equal to a fraction, which is slightly different for each of the functions:

$$\sin\theta = \frac{\text{opposite side}}{\text{hypotenuse}} \qquad \cos\theta = \frac{\text{adjacent side}}{\text{hypotenuse}} \qquad \tan\theta = \frac{\text{opposite side}}{\text{adjacent side}}$$

Let me explain what the terms in these ratios mean. In Figure 10.4, I've drawn two copies of a right triangle ABC; the only thing that differs about the two triangles is the acute angle I'm choosing to focus on at the moment (labeled θ).

Figure 10.4

Though both of these triangles are exactly the same, each version focuses on one of the two acute angles. Notice that the labels "adjacent" and "opposite" vary depending upon the acute angle involved, but the hypotenuse always stays the same.

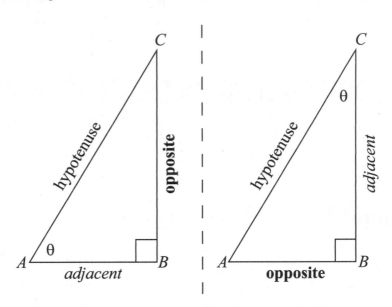

Both right triangles have one distinct side that is the longest $\left(\overline{AC}\right)$, called the *hypotenuse*. It's across from the right angle ($\angle B$) and acts as one side of each acute angle.

Talk the Talk

Each acute angle in a right triangle has two distinct sides: the **adjacent side** and the **hypotenuse** (which is the longer of the two). The side of a right triangle that does not form the acute angle in question is called the **opposite side**.

The nonhypotenuse segment that completes the acute angle is called the *adjacent side*. In the triangle on the right in Figure 10.4, θ marks $\angle C$ as the acute angle I'm interested in. Notice that $\angle C$ is made up of two segments: $\overline{AC}$ (the hypotenuse) and $\overline{BC}$ (which completes the angle and is, therefore, the adjacent side). The remaining side of the right triangle, $\overline{AB}$, is called the *opposite side*. In the left triangle in Figure 10.4, the opposite and adjacent sides switch places as you focus on the other acute angle.

To figure out the sine, cosine, or tangent of an acute angle in a right triangle, all you have to do is figure out which side is which, and then plug their lengths into the appropriate ratio. You'll use your calculator to evaluate almost every trigonometric function, with the exception of a small handful that belong to something called the "unit circle"; I'll describe that at the end of the chapter.

Example 3: Calculate the length of x in Figure 10.5 and round your answer to the thousandths.

Kelley's Cautions

When you use your calculator to evaluate trig ratios, it needs to be set in the correct mode (either degrees or radians), based on how the angle is measured. For instance, sin 45° ≈ 0.707106781 if your calculator is correctly set in degrees mode, but if you're in radians mode, you get an incorrect value of 0.85090352.

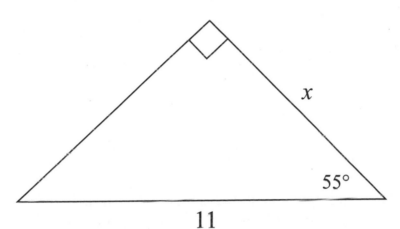

Figure 10.5

To figure out what x is, you first need to figure out which trigonometric ratio to use.

The side with length 11 is opposite the right angle in the diagram, so it must be the hypotenuse; it's also one side of the 55° angle. The other side of the angle (the adjacent side) is the value you're trying to solve for: x. So, you know the hypotenuse length, and you're trying to find the adjacent side length. Which trig ratio governs the relationship between hypotenuse and adjacent sides? Cosine does, so plug in the 55° angle and the sides that correspond to it:

$$\cos\theta = \frac{\text{adjacent}}{\text{hypotenuse}}$$

$$\cos 55° = \frac{x}{11}$$

Multiply both sides of the equation by 11 to solve for x. Because the problem asks you to express your answer as a decimal, type cos 55 into your calculator, ensuring that it's set in degrees mode, and you'll find that cos 55° ≈ 0.573576436.

$$x = 11(\cos 55°)$$
$$x ≈ 6.309$$

You've Got Problems

Problem 3: Calculate x in Figure 10.6, and round your answer to the thousandths.

Figure 10.6

Is there anything better than solving for x? *What a rush!*

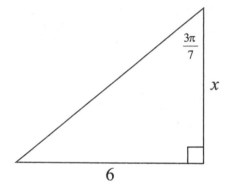

The Unit Circle

The three trigonometric functions you've learned so far (sine, cosine, and tangent) are even more useful than I may have led you to believe. You may think that they work only for acute angles of right triangles, but that's not true. In fact, every single angle in the universe has a sine and cosine value, and nearly every angle has a tangent as well. However, to get the most use out of them, you need to slap them onto the coordinate plane in standard position.

In Figure 10.7, I've drawn something called the *unit circle*. The origin of the name is simple—it's a circle centered at the origin whose radius is 1 unit. (It doesn't matter whether the "unit" in question is centimeters, inches, feet, miles, kilometers, light-years, corncobs, fingernails, or bacon strips—in case you're watching your carbohydrate intake.)

Talk the Talk

The **unit circle**, though merely a circle centered at the origin with a radius of 1, is a tool to generate the cosine and sine values of the most common angles encountered in precalculus. If an angle θ is drawn in standard position, it will intersect the unit circle at a point (a,b) where $a = \cos θ$ and $b = \sin θ$.

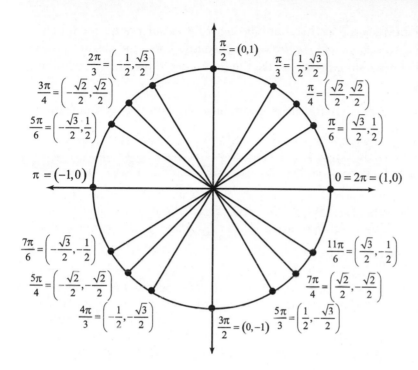

Figure 10.7

The x- and y-values of the point where an angle's terminal side hits the unit circle give you the cosine and sine values of that angle.

Besides the circle, Figure 10.7 features a humongous pile of angles drawn in standard position: 0 (0°), $\frac{\pi}{6}$ (30°), $\frac{\pi}{4}$ (45°), $\frac{\pi}{3}$ (60°), $\frac{\pi}{2}$ (90°), and so on, until you get back to the positive x-axis after one full counterclockwise revolution and end up at 2π (360°). Believe it or not, the coordinate pairs marking the intersection of each angle's terminal side and the unit circle are incredibly important, so I have included those points as well.

Those points are so essential because their x-values are the cosines of the angles intersecting there, and their y-values are the sines. For instance, the angle $\frac{11\pi}{6}$ intersects the unit circle at the point $\left(\frac{\sqrt{3}}{2}, -\frac{1}{2}\right)$, so you automatically know that $\cos\frac{11\pi}{6} = \frac{\sqrt{3}}{2}$ and $\sin\frac{11\pi}{6} = -\frac{1}{2}$. In other words, all you have to do to instantly remember the cosine and sine values of 17 important angles is to memorize this measly little unit circle illustration (or get it tattooed on your arm, which is both more permanent and more painful).

Critical Point

You'll actually know far more than 17 angles when you memorize the unit circle. Even though it covers angles only between 0 and 2π, *coterminal angles will have equal trigonometric values,* so you actually are memorizing information for an infinite number of angles!

Sure, Figure 10.7 is intimidating at first, but burning it into your brain is definitely worth it. Most calculus teachers will *require* you to instantly know the cosine and sine values of every angle on the unit circle, so take the time now to memorize it. Finding patterns there can help:

◆ Fractions with the same denominators always have the same cosines and sines; they differ only by sign.

How'd You Do That?

You can figure out the intersection points on the unit circle yourself, if you really feel like it. You just need to remember the 45°-45°-90° and 30°-60°-90° right triangle relationships you learned in geometry. As an example, I'll show you how to generate the point $\left(\frac{\sqrt{3}}{2}, \frac{1}{2}\right)$, where the angle $\frac{\pi}{6}$ hits the unit circle.

To begin, I'll redraw the angle in standard position, labeling the points for easy reference, and connecting the end of the terminal side (*B*) to the *x*-axis, where it will meet at point C (see Figure 10.8). This creates a right triangle ABC, whose hypotenuse, $\overline{AB}$, has length 1 because it is also a radius of the unit circle. You also know that $m\angle B = 60°$, since all the angles in a triangle must add up to 180°, and the other two angles sum up to 120°.

Figure 10.8

The angle $\frac{\pi}{6}$ equals 30°, and rewriting it like that might help trigger your deep, dark memories of 30°-60°-90° triangles from geometry.

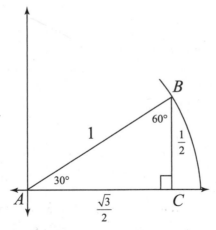

According to the 30°-60°-90° theorem of geometry, the side opposite the 30° angle $\left(\overline{BC}\right)$ must have a length of $\frac{1}{2}$ times the hypotenuse, so $BC = \frac{1}{2} \cdot 1 = \frac{1}{2}$. By the same theorem, the side opposite the 60° angle $\left(\overline{AC}\right)$ must be $\sqrt{3}$ times as long as $\overline{BC}$, so $\overline{AC} = \sqrt{3} \; \frac{1}{2} = \frac{\sqrt{3}}{2}$.

♦ Angles in the same quadrant have the same sign pattern: I = (+,+), II = (−,+), III = (−,−), and IV = (+,−)

Trust me, although you may feel uncomfortable around it at first, the unit circle will soon become your best friend. Learn it. Memorize it. Take it to your favorite restaurant and chat about the meaning of life, and whether or not you think that Alf puppet is funny.

Example 4: Using the unit circle and coterminal angles but no calculator, calculate $\cos\left(-\dfrac{\pi}{2}\right)$.

Solution: Find an angle coterminal to $-\dfrac{\pi}{2}$ that falls on the unit circle by adding 2π to it:

$$-\frac{\pi}{2} + 2\pi = -\frac{\pi}{2} + \frac{4\pi}{2} = \frac{3\pi}{2}$$

Since $-\dfrac{\pi}{2}$ and $\dfrac{3\pi}{2}$ are coterminal, $\cos\left(-\dfrac{\pi}{2}\right) = \cos\left(\dfrac{3\pi}{2}\right)$ because coterminal angles have the same trig values. Notice that the angle $\dfrac{3\pi}{2}$ intersects the unit circle at the point $(0,-1)$; the cosine is the x-value of that coordinate pair, so $\cos\left(\dfrac{3\pi}{2}\right) = 0$.

You've Got Problems

Problem 4: Using the unit circle and coterminal angles but no calculator, calculate $\sin\dfrac{21\pi}{4}$.

The Least You Need to Know

♦ An angle in standard position has a vertex at $(0,0)$, and its initial side overlaps the positive x-axis.

♦ Multiply a degree measurement by $\dfrac{\pi}{180}$ to convert it to radians, and multiply a radian measurement by $\dfrac{180}{\pi}$ to convert it to degrees.

♦ Adding 2π (or 360°) to or subtracting it from an angle in standard position results in a coterminal angle, which will have the same trigonometric function values.

♦ The x-coordinate of an angle's intersection point with the unit circle is that angle's cosine value; the y-coordinate is its sine value.

Chapter 11

Graphing Trigonometric Functions

In This Chapter

- The infinite loop of periodic functions
- Graphing sine and cosine
- Applying transformations to trigonometric graphs
- Defining and graphing tangent, cotangent, secant, and cosecant

When I was 10 (and even when I was 20), I had those really crazy kid aspirations. You know what I mean—I wanted to be an astronaut businessman who becomes president as he climbs Mt. Everest during his summers off as a secret agent superspy. Plus, I'd be in a rock-and-roll band (despite the fact that I could play only the trumpet, arguably not the most rockin' instrument in the land), and every woman in the world would be mad with desire at the very sight of me. Then you turn 30, and you realize things aren't quite the way your wildest dreams predicted.

Instead of a sexy presidential James Bond, I am turning into a balding Pillsbury Dough Boy who enjoys math more than moon exploration and whose most exciting athletic activity is a rousing game of computer football.

Before long, life is somewhat predictable. I get up, shower, go to work, try to find someplace to eat lunch for less than $10 (good luck in Washington, D.C.), drive home, hang with the family, do the dishes, hose off the kid, watch a little reality television with the wife, hit the sack, and then start over again tomorrow.

Every once in a while I wonder, "Am I in a rut? How come I'm not exploring Mars or trying unsuccessfully to fend off aggressive sexual advances from Rebecca Romijn?" When I was 10, I thought I had unlimited potential and that every day would be an adventure; it turns out that the biggest adventure I usually have is when I misplace my car keys. But that's the funny thing—whenever my "rut" gets disturbed, it makes me unhappy. A predictable life for me is a very happy life, much to the chagrin of my 10-year-old self. Maturity brings appreciation of the things I once feared.

The same thing goes with math functions. (I know, you doubt that I can tie this back to math, but just stand back and watch the magic happen.) Just about all of the functions we've discussed so far (like exponential, radical, rational, and polynomial functions) shoot off to infinity (or down to negative infinity) at the edges of their graphs. They have high hopes and are aiming for the stars in their simplistic designs on life. Trigonometric functions, as you'll see in this chapter, have much more reasonable goals—they have a happy rut and would rather repeat their graphs over and over for all of eternity, like me.

Periodic Functions

All six trig functions (right now, you know only three of them—sine, cosine, and tangent—but you'll know the rest by the end of the chapter) are *periodic functions*. Like the metaphor of my happy and predictable life, each has its highs and lows but will ultimately repeat itself over and over for all time. Check out the periodic function I've drawn in Figure 11.1.

A quick glance at the function in Figure 11.1 should be all it takes to figure out that it is periodic because it repeats itself over and over like a crazed heart monitor. One key question you must ask yourself with periodic functions is this: "How long (counted in horizontal units on the coordinate plane) does it take for the graph to repeat itself?" The answer to that question is a number called the *period* of the function.

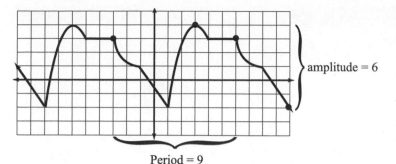

amplitude = 6

Period = 9

Figure 11.1

A periodic function can be characterized by its period and amplitude, two attributes that can be counted out on a coordinate plane or calculated by other clever, time-saving, and less boring techniques.

To calculate the period, identify two corresponding points on two consecutive repetitions of the graph. In Figure 11.1, I've marked (–3,3) and (6,3); they're easy to spot because they mark the end of the horizontal section of each repeated segment. Count the number of horizontal spaces you have to travel to get from (–3,3) to (6,3), and you'll get a period of 9.

If you'd rather use a more formulaic technique to calculate the period, subtract the x-values of the points and take the absolute value: $|-3-6| = 9$ or $|6-(-3)| = 9$. (The y-values of the points you choose will always be equal because once you finish a full repetition of the graph, you'll be back to the height at which you started.)

Talk the Talk

A **periodic function** repeats itself after a fixed horizontal distance called its **period** The total height of a periodic function divided by 2 is called the **amplitude**.

The *amplitude* of a periodic function describes the span between the middle of the function and its highest or lowest point. In other words, it measures half the total height of the function. To calculate it, first identify one each of the highest and lowest points on the graph—I've marked (3,4) and (10,–2)—and either count the number of vertical spaces between them or calculate the absolute value of the difference of their y-values. Either way, you'll end up with 6. To get the amplitude, divide that number by 2: $6 \div 2 = 3$.

You've Got Problems

Problem 1: Identify the period and amplitude of the following periodic function based on its graph:

Figure 11.2

Few of the periodic functions from here on will be so sharp and pointy, so feel free to remove your safety goggles.

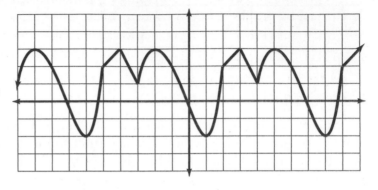

Graphing Sine and Cosine

I already spoiled the surprise and told you that all of the trigonometric functions were periodic, but can you figure out why? What is it about the nature of sine and cosine that suggests their values will continuously repeat themselves? Remember, the values of those two functions come from the unit circle, which I discussed at the end of Chapter 10. Once angles go over 2π, they have looped all the way around the unit circle and start repeating themselves as coterminal angles.

Hold on a minute, let me say that again: Once the angles of sine and cosine get bigger than 2π, the functions start repeating themselves. Sounds familiar, doesn't it? That means 2π is the period for sine and cosine! It'll be easier to verify this, however, if you look at a graph of sine and treat it like the periodic functions you examined a few pages ago.

CAUTION **Kelley's Cautions**

Although trigonometric angles are usually expressed as Greek letters, it's not uncommon to see functions written as "$f(x) = \sin x$" rather than "$f(\theta) = \sin \theta$." Sometimes I'll write the functions in terms of x, and other times I will use θ, but whichever variable is used, it doesn't change how you'll do the problems.

Connect the Dots for Fun and Profit

To graph $f(x)$ = sin x, I'll plot a bunch of points, until I can get a good idea of the graph's shape. I'll warn you, though, that graphs of trigonometric functions look a little different from the graphs of lines and polynomials. The biggest change is the way you'll work with the x-axis. Since trig functions require angle inputs, the x-axis will have angles from the unit circle (like $\frac{\pi}{4}, \frac{\pi}{2}$, and π) rather than the integers you're used to.

To get the most accurate graph I can, I'll use all of the angles in the unit circle between $\theta = 0$ and $\theta = 2\pi$ to generate the graph of $f(x)$ = sin x. No real computation is involved, since the y-value of each unit circle point represents the sine. Therefore, $\sin(0) = 0$, $\sin\left(\frac{\pi}{6}\right) = \frac{1}{2} = 0.5$, $\sin\left(\frac{\pi}{4}\right) = \frac{\sqrt{2}}{2} \approx 0.707$, $\sin\left(\frac{\pi}{3}\right) = \frac{\sqrt{3}}{2} \approx 0.866$, $\sin\left(\frac{\pi}{2}\right) = 1$, and so on, all the way to sin $(2\pi) = 0$.

That means the points $(0,0)$, $\left(\frac{\pi}{6}, 0.5\right)$, $\left(\frac{\pi}{4}, 0.707\right)$, $\left(\frac{\pi}{3}, 0.866\right)$, $\left(\frac{\pi}{2}, 1\right)$, all the way up to $(2\pi, 0)$ make up the graph of $f(x)$ = sin x, as you can see in Figure 11.3.

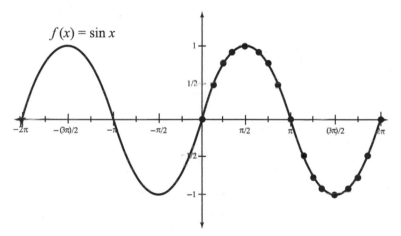

Figure 11.3

The graph of sin x is often referred to as the sine wave and even has its own adjective—any graph that has a wavy sinelike shape is described as sinusoidal (which actually sounds like a good name for an allergy medicine).

Critical Point

If you were to plot the number of daylight hours there are at your home each day over the course of a year, you'd end up with a graph that looks almost *exactly* like y = sin x.

In this analogy, $\theta = 0 = 2\pi$ represents the spring equinox, and $\theta = \pi$ represents the winter equinox (the hours of daylight and darkness are exactly equal). The summer and winter solstices (the most and fewest hours of daylight, respectively) would occur at $\theta = \frac{\pi}{2}$ and $\theta = \frac{3\pi}{2}$.

All of the points in Figure 11.3 mark the angles of the unit circle. (Since sine is periodic, I went ahead and drew another period of the graph on the interval $[-2\pi,0]$ as well.) Now that you've got the graph (and can see that it definitely falls into the periodic category), verifying the period and amplitude are a snap. The origin and $(2\pi,0)$ represent corresponding points on consecutive repetitions of the graph, so the period is $|0-2\pi|=2\pi$, just like we guessed it would be. To get the amplitude, subtract one of the lowest heights (–1) from one of the highest (1). Take the absolute value $\left(|1-(-1)|=2\right)$ and divide by 2 to get an amplitude of $2 \div 2 = 1$.

The sine graph also has these important attributes:

Critical Point _____

Remember, the way to mathematically state "all the coterminal angles" of some angle θ is to write "θ + 2*k*π, where *k* is an integer," since adding and subtracting a multiple of 2π is how you actually generate coterminal angles.

♦ All *x*-intercepts fall at integer multiples of π: ..., $-2\pi, -\pi, 0\pi, \pi, 2\pi$, ... In other words, the angles $\theta = 0$, $\theta = \pi$, and all their coterminal angles ($\theta = 0 + 2k\pi$ and $\theta = \pi + 2k\pi$), where *k* is an integer.

♦ The maximum value of the sine graph (1) occurs at $\theta = \frac{\pi}{2}$ and all its coterminal angles: $\theta = \frac{\pi}{2} + 2k\pi$; the minimum value (–1) occurs at $\theta = \frac{3\pi}{2}$ and all its coterminal angles $\left(\theta = \frac{3\pi}{2} + 2k\pi\right)$.

♦ The *domain* of $f(x) = \sin x$ is all real numbers, and the *range* is $[-1,1]$.

Talk the Talk _____

Just in case you forgot, the **domain** of a function is the collection of its valid inputs; the **range** is the set of numbers that are possible outputs. For instance, the domain of $f(x) = \sqrt{x}$ is $[0,\infty)$ (you can take the square root of only 0 or a positive number), and the range is $[0,\infty)$ as well (when you take the square root of something, you'll get either 0 or a positive number).

Add the graph of sine to your memorized repertoire. Very soon, I'll start asking you to graph sine functions with transformations, and it'll save you a lot of time (and reduce your frustrated weeping considerably) if you know the graph by heart.

Cosine's Time to Shine

You can use the same technique I used for sine to get the graph of $g(x) = \cos x$ (Figure 11.4). While it's no big surprise that cosine also has a period of 2π and an amplitude of 1 (it comes from the unit circle, just like sine, and has all the same values, though in a different order), you might be surprised to see that it has the same shape. In fact, if you were to shift the graph of cosine to the right $\frac{\pi}{2}$ units, you'd get the *exact* graph of sine.

$g(x) = \cos x$

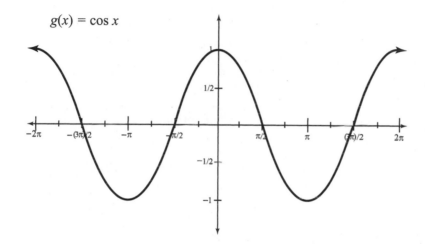

Figure 11.4

The graph of g(x) = *cos* x *is eerily reminiscent of a sine graph. It's almost like they're identical twins (but you can tell them apart because sine has a little mole on its origin).*

The major features of cosine graph mirror those of sine. It has repetitive x-intercepts (that occur at $\theta = \frac{\pi}{2} + 2k\pi$ and $\theta = \frac{3\pi}{2} + 2k\pi$), a maximum value of 1 (at $\theta = 0 + 2k\pi$), and a minimum value of -1 (at $\theta = \pi + 2k\pi$).

Graphing Trig Transformations

The function transformations you learned in Chapter 6 work the same way when you're graphing trigonometric functions. Adding to a trig function still moves the function up, and subtracting inside the function itself still moves the function right. You should know two tips, however, that are especially useful for periodic functions because they deal with the amplitude and period:

◆ **If sine or cosine is multiplied by a number, the absolute value of that number is the amplitude.** In the function $k(x) = -3\cos x + 1$, the -3 out front tells you that the amplitude of the graph is $|-3| = 3$. Practically, this means that the graph of $k(x)$ (before it is moved up one unit) will stretch to a maximum

height of 3 and a minimum height of –3, instead of the usual maximum cosine height of 1 and minimum height of –1.

♦ **The coefficient of x tells you how many times the graph will repeat itself in the original period of the function.** In the function $j(x) = 4\sin(2x) - 1$, the coefficient of x is 2, which means that two full repetitions of the graph will occur on $[0, 2\pi]$ (the original period), rather than just 1. To calculate the period of $j(x)$, use this formula:

$$\text{new period} = \frac{\text{original period}}{x\text{-coefficient}}$$

Since the x-coefficient of $j(x)$ is 2 and the original period of sin x is 2π, the period of $j(x)$ is $\frac{2\pi}{2} = \pi$.

Remember, multiplying inside a function by a number greater than 1 causes the function to get squeezed horizontally toward the y-axis, and that's exactly what happens here. The amplitude of the graph doesn't change; $j(x)$ just swoops up and down quickly, twice as fast as it would without that 2 coefficient. (You'll graph $j(x)$ in the next "You've Got Problems" sidebar.)

Trigonometric graph transformations are nothing to stress over. Put most of your effort into drawing one, good, accurate period of the graph on the coordinate plane, and then just fill up the rest of the horizontal axis with copies of that period.

Example 1: Sketch the graph of $k(x) = -3\cos x + 1$ on the interval $[-2\pi, 2\pi]$.

Solution: This problem specifically asks you to draw a graph whose x-axis begins at -2π and ends at 2π, but you should always assume that this is the interval you should graph if none is indicated by the problem. (Notice that Figures 11.3 and 11.4 both use this interval as well.)

Three transformations change plain old $\cos x$ into $k(x)$: multiplying the function by 3 (which makes its amplitude 3), applying a reflection across the x-axis (since the 3 is negative), and vertically shifting the entire graph up one unit. You should apply those transformations in that order, and you'll end up with the graph in Figure 11.5.

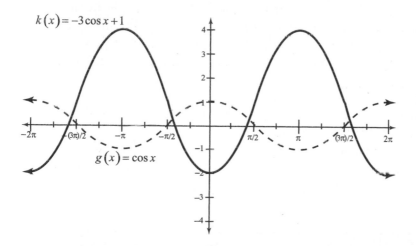

Figure 11.5

The graph of k(x) = −3cos x + 1 *(the solid curve) and the original, untransformed graph of* y = cos x *(the dotted curve).*

You've Got Problems

Problem 2: Sketch the graph of $j(x) = 4\sin(2x) - 1$ on the interval $[-2\pi, 2\pi]$.

Meet the Rest of the Family

At this point, you're well into trigonometry, and it's too late to try to turn back. As they say in poker (which I find strangely hypnotic on television), you're "pot dedicated"—you've put so much into it now that it wouldn't be a wise investment of resources to give up and quit.

It's sort of like dating the same person for so long that you're required to go meet the rest of his or her family. Sure, you might have met a brother or sister or even a parent already, but once you've crossed that indescribable threshold of dating "long enough," you have to put in some time meeting the aunts, uncles, cousins, grandparents, sister's friend's favorite hairstylist's neighbor's cat, and so on.

So put a big smile on your face (even if it's fake), and get ready to meet the rest of the trig functions. Make sure to learn all of their graphs while you're at it.

Tangent

You've already met tangent, but you don't know a heck of a lot about it. In fact, you'd probably be surprised to find out that not only is the tangent equal to the opposite side divided by the adjacent side of an acute angle (like you learned in the section "Right Triangle Trigonometry" in Chapter 10), but it's always equal to the sine of an angle divided by its cosine: $\tan\theta = \dfrac{\sin\theta}{\cos\theta}$.

The graph of tangent (see Figure 11.6) is quite different than the graphs of sine and cosine, even though its values come directly from their quotient. This is because tangent is a rational function—it's got $\cos\theta$ in the denominator, which has a value of 0 at $\theta = \dfrac{\pi}{2}$, $\theta = \dfrac{3\pi}{2}$, and all of their coterminal angles. Therefore, tangent will have a vertical asymptote at every one of those locations.

Figure 11.6

The graph of y = *tan* x. *Notice that the tangent function passes through the origin (since* $\tan 0 = \dfrac{\sin 0}{\cos 0} = \dfrac{0}{1} = 0$ *) and has asymptotes at all the* $\dfrac{k\pi}{2}$ *'s, where* k *is an odd integer.*

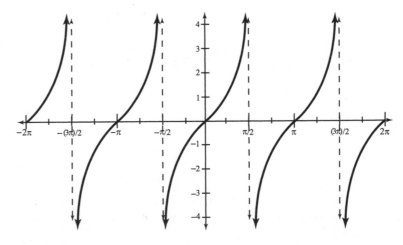

The other major differences between tangent's graph and those of sine and cosine is that tangent has a period of π (instead of 2π) and doesn't have an amplitude. Since tangent's graph extends infinitely high and low, there are no "maximum" and "minimum" heights, so it would be impossible to calculate the absolute value of their difference. (In fact, none of the remaining trig functions have a measurable amplitude like sine and cosine.)

Cotangent

The reciprocal of $\tan \theta$ $\left(\dfrac{\cos\theta}{\sin\theta}\right)$ is known as the cotangent function, and it's abbreviated cot θ. Cotangent is also called the *cofunction* of tangent because it has the same name as the tangent function, but with the prefix "co-" stapled on the front. Like tangent, the graph of cotangent (see Figure 11.7) will have a period of π and a whole bunch of vertical asymptotes, this time located at every multiple of π, where its denominator ($\sin \theta$) equals 0.

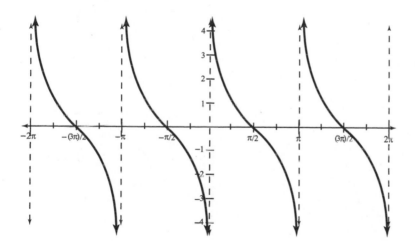

Figure 11.7

The cotangent graph looks like the tangent graph, just reflected about the y-axis (it goes up to the left and down to the right, the opposite of tangent) and moved $\dfrac{\pi}{2}$ units to the right.

It may have taken a while to figure out what the sine, cosine, and tangent were because they were all defined as ratios of the sides of right triangles. However, cotangent couldn't be easier: All you do is take the reciprocal of tangent. That's all there is to it.

Talk the Talk

If two functions have the same name as one another, except that one starts with "co-" (like sine and **co**sine or tangent and **co**tangent), they are called (appropriately enough) **cofunctions**.

Secant and Cosecant

The final two trigonometric functions are also just reciprocals of trig functions you already know. Secant (abbreviated "sec") is defined as the reciprocal of cosine

Kelley's Cautions

Secant and sine are *not* reciprocals of one another (neither are cosecant and cosine), even though they start with the same letter, so don't let that confuse you.

$\left(\sec\theta = \dfrac{1}{\cos\theta}\right)$, and cosecant (abbreviated "csc") is the reciprocal of sine $\left(\csc\theta = \dfrac{1}{\sin\theta}\right)$. Like their reciprocals, secant and cosecant have periods of 2π, but because they're rational functions, the graphs of secant and cosecant will once again feature vertical asymptotes and infinitely tall curves.

Secant's graph (see Figure 11.8) will have the same vertical asymptotes as tangent because it shares the same denominator ($\cos\theta$).

Figure 11.8

The graph of y = sec θ— *an infinite pattern of upward- and downward-facing humps—is the solid curve on the coordinate axes. The dotted curve is the graph of cosine (the reciprocal of secant), which helps marks the minimum and maximum points on secant's graph and has x-intercepts at every vertical asymptote.*

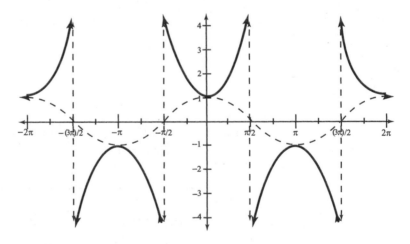

To draw the graph of secant, first toss in the vertical asymptotes where they belong (at $\dfrac{\pi}{2}$, $\dfrac{3\pi}{2}$, and coterminal angles), and then lightly sketch the graph of cosine. The bowl-shape curves of secant will blossom out of cosine's maximum and minimum points, quickly nearing (but never crossing) the vertical asymptotes.

The graph of cosecant is very similar to secant. As you can see in Figure 11.9, it has vertical asymptotes (this time, at every multiple of π) and bowl-shape plumes that grow out of sine's maximum and minimum points. Notice that the range for both secant and cosecant is $(-\infty,-1]$ or $[1,\infty)$—only the function values between -1 and 1 are excluded.

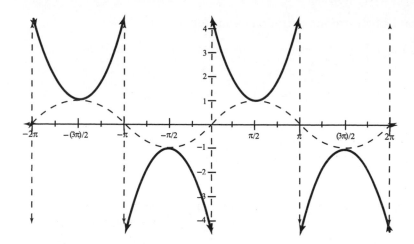

Figure 11.9

This time, the dotted curve is the graph of y = sin θ, *and sprouting from the highest and lowest points of the dotted graph is its reciprocal* (y = csc θ), *the solid curve.*

Critical Point

Without a single calculation, you can instantly figure out whether any trig function for a given angle is going to be positive or negative. All you need to know is which quadrant the angle's terminal side lies in.

Figure 11.10

The functions listed in each quadrant tell you what trig functions are positive for terminal sides located there.

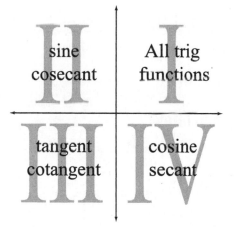

In Figure 11.10, I've labeled each quadrant with the trig functions that are positive if an angle's terminal side is located there. For instance, the terminal side of $\theta = \frac{5\pi}{3}$ lies in quadrant IV, so you automatically know that $\cos\frac{5\pi}{3}$ and $\sec\frac{5\pi}{3}$ are positive, but $\sin\frac{5\pi}{3}$, $\tan\frac{5\pi}{3}$, $\cot\frac{5\pi}{3}$, and $\csc\frac{5\pi}{3}$ will all be negative.

Example 2: Evaluate all six trigonometric functions for $\theta = \frac{3\pi}{4}$.

Solution: The cosine and sine values for θ come straight from the unit circle: $\cos\theta = -\frac{\sqrt{2}}{2}$ and $\sin\theta = \frac{\sqrt{2}}{2}$. Tangent is equal to the quotient of the sine and cosine values, but since they're opposites of one another, they're going to cancel out, leaving a tangent value of -1:

$$\tan\frac{3\pi}{4} = \frac{\sin\frac{3\pi}{4}}{\cos\frac{3\pi}{4}} = \frac{\frac{\sqrt{2}}{2}}{-\frac{\sqrt{2}}{2}} = -1$$

Calculating the remaining three functions is simple; they're the reciprocals of the functions you just figured out. Cotangent is the reciprocal of tangent (-1), secant is the reciprocal of cosine $\left(\frac{-2}{\sqrt{2}}\right)$, and cosecant is the reciprocal of sine $\left(\frac{2}{\sqrt{2}}\right)$. Some math teachers get cranky when they see radical signs in the denominator of a fraction and would rather you multiply the top and bottom of secant and cosecant by $\sqrt{2}$ to rewrite the fractions. That's not absolutely necessary, but if you do, you'll get $\sec\frac{3\pi}{4} = -\frac{2}{\sqrt{2}} \cdot \frac{\sqrt{2}}{\sqrt{2}} = -\frac{2\sqrt{2}}{2} = -\sqrt{2}$; similarly, $\csc\frac{3\pi}{4} = \sqrt{2}$.

You've Got Problems

Problem 3: Evaluate all six trigonometric functions for the angle $\theta = \frac{5\pi}{6}$.

The Least You Need to Know

- Periodic functions repeat themselves every time they pass a fixed horizontal distance called the period.
- All six trigonometric functions (sine, cosine, tangent, cotangent, secant, and cosecant) are periodic.
- Cotangent is the reciprocal of tangent, secant is the reciprocal of cosine, and cosecant is the reciprocal of sine.
- Cosine, sine, secant, and cosecant have a period of 2π, but tangent and cotangent have a period of π.

12

Trigonometric Identities

In This Chapter

♦ Describing the difference between equations and identities

♦ Highlighting basic identities

♦ Simplifying expressions using identities

♦ Verifying identity equations

♦ Applying double-angle identities

One of the big buzzwords that has arisen at the beginning of the twenty-first century is "identity theft"—a crime in which some jerk figures out some of your most personal and uniquely identifying information (like your birthday, Social Security number, mother's maiden name, shortest cousin, and least favorite Sesame Street character) and uses it to pose as you. Maybe these thieves apply for a credit card with your name on it and snatch it out of your mailbox before you get home (racking up debt like the world's going to end), or maybe they just steal your credit card number outright, and before you know it, MasterCard is calling and they want to know why you spent $14,000 on penguins.

Of course, you never bought penguins (and if you have, I'd rather not know about it, to be honest), so you rant and rave until you figure out that someone's out there masquerading as you and, for some reason, buying exotic birds. Where will the madness end? Owls? Toucans? Pterodactyls? You feel violated and angered by all these foul purchases (you saw that horrible pun coming, didn't you?), and the credit card company's fraud department swings into action.

Like hawks (again with the birds), these accountant superheroes watch the activity on your card closely to see who has your identity, even though the perp (as they say in the biz) probably won't look anything like you. In this chapter, you play the part of the credit company fraud inspector (it would help if you could find something tweed to wear), proving that two trigonometric statements, although bearing no resemblance whatsoever, actually have the exact same identity.

Basic Training

Before criminals (and trigonometric equations) fear you, some training is required. You'll have to learn the most common ways trig expressions masquerade as one another. This section focuses on the most basic *identities*. Your first lesson: What the heck is an identity? Well, it's an equation, but a special kind of equation—whereas most equations have only one, two, or a fixed finite number of solutions, an identity is true no matter what value you plug into it (as long as the value is legal—if it makes one of the functions undefined, all bets are off).

Talk the Talk

A trigonometric **identity** is an equation that is true no matter what angle is substituted into it.

According to that definition, then, this equation is actually an identity:

$$\cos \theta + \sin \theta = \sin \theta + \cos \theta$$

Of course, the equation's true—the order doesn't matter when you add two things—but you already knew that way back in Chapter 1 because that's just the commutative property. The identities I'm about to introduce you to are slightly different than this lame example because they're not obviously equal at first glance, which will make them more powerful tools later.

Sign Identities

Different trig functions act in different ways when you insert a negative input, and that's the basis of the sign identities:

♦ $\cos (-\theta) = \cos \theta$

♦ $\sec (-\theta) = \sec \theta$

◆ $\sin(-\theta) = -\sin\theta$

◆ $\tan(-\theta) = -\tan\theta$

◆ $\cot(-\theta) = -\cot\theta$

◆ $\csc(-\theta) = -\csc\theta$

Basically, this means that the cosine and secant values of an angle and its opposite are the exact same thing, so $\cos\left(\dfrac{\pi}{4}\right)$ and $\cos\left(-\dfrac{\pi}{4}\right)$ are equal, as are their reciprocals, the secant values. However, the four other trig functions will give opposite values when you plug in opposite angles. For example, $\sin\left(\dfrac{\pi}{2}\right) = 1$, but $\sin\left(-\dfrac{\pi}{2}\right) = -1$.

How'd You Do That?

Only cosine and secant give the same output for opposite inputs because they are even functions. In algebra, you learned that every point (x,y) on an even function has a sister point $(-x,y)$ on the graph, which not only ensures that opposite inputs have the same outputs, but actually causes the graph to be y-symmetric. (The pieces of the graph on either side of the y-axis are exact reflections of one another.)

Sine, tangent, cosecant, and cotangent are all odd functions, with graphs symmetric about the origin. In these four functions, if you've got a point (x,y) on the graph, you'll also have the point $(-x,-y)$, so opposite inputs (x and $-x$) result in opposite outputs (y and $-y$).

Cofunction Identities

Back when I was introducing the cotangent function in Chapter 11, I explained that its graph looked a lot like tangent's. Specifically, if you reflect the tangent graph about the y-axis and then move it $\dfrac{\pi}{2}$ units to the right, you get the cotangent graph. Amazingly enough, if you do the same thing to the cotangent graph, you'll end up with the tangent graph again.

This strange property is not specific to just the tangent and cotangent functions. In fact, those two transformations turn any trig function into its cofunction, which is the basis for the six cofunction identities:

- $\cos\left(\dfrac{\pi}{2} - \theta\right) = \sin\theta$ and $\sin\left(\dfrac{\pi}{2} - \theta\right) = \cos\theta$

- $\tan\left(\dfrac{\pi}{2} - \theta\right) = \cot\theta$ and $\cot\left(\dfrac{\pi}{2} - \theta\right) = \tan\theta$

- $\sec\left(\dfrac{\pi}{2} - \theta\right) = \csc\theta$ and $\csc\left(\dfrac{\pi}{2} - \theta\right) = \sec\theta$

How'd You Do That?

To shift a function, like $f(\theta) = \cot\theta$, $\dfrac{\pi}{2}$ units to the right, you subtract that value inside the function: $f(\theta) = \cot\left(\theta - \dfrac{\pi}{2}\right)$. If you want to reflect $f(\theta)$ across the y-axis, multiply the inside of the function by a negative: $f(\theta) = \cot\left(-\theta + \dfrac{\pi}{2}\right)$, which is the same as $f(\theta) = \cot\left(\dfrac{\pi}{2} - \theta\right)$. That's where the quantity in parentheses comes from for the cofunction identities.

Example 1: Demonstrate the cofunction identity $\sin\left(\dfrac{\pi}{2} - \theta\right) = \cos\theta$ if $\theta = \dfrac{\pi}{4}$.

Solution: If you plug $\theta = \dfrac{\pi}{4}$ into the identity, you should get a true result:

$$\sin\left(\frac{\pi}{2} - \frac{\pi}{4}\right) = \cos\frac{\pi}{4}$$

$$\sin\left(\frac{2\pi}{4} - \frac{\pi}{4}\right) = \cos\frac{\pi}{4}$$

$$\sin\frac{\pi}{4} = \cos\frac{\pi}{4}$$

$$\frac{\sqrt{2}}{2} = \frac{\sqrt{2}}{2}$$

Reciprocal Identities

In Chapter 11, you learned that cotangent, secant, and cosecant are defined as the reciprocals of tangent, cosine, and sine (in that order), so these identities are nothing new.

- $\sec\theta = \dfrac{1}{\cos\theta}$ and $\cos\theta = \dfrac{1}{\sec\theta}$

- $\csc\theta = \dfrac{1}{\sin\theta}$ and $\sin\theta = \dfrac{1}{\csc\theta}$

- $\tan\theta = \dfrac{1}{\cot\theta}$ and $\cot\theta = \dfrac{1}{\tan\theta}$

Keep in mind that the reciprocal relationship works both ways, so if cotangent is the reciprocal of tangent, tangent is the reciprocal of cotangent as well.

Pythagorean Identities

In geometry (if not before that), you learned the Pythagorean Theorem, which states that a right triangle with legs a and b and hypotenuse c will satisfy the equation $a^2 + b^2 = c^2$. Since the values of sine and cosine are based on a right triangle embedded in a unit circle, you can use the Pythagorean Theorem to create one of the most useful trigonometric identities.

In Figure 12.1, I've drawn a generic angle (θ) on the unit circle. Remember, the intersection point of its terminal ray and the unit circle will be ($\cos \theta$, $\sin \theta$). Think about what that means: To get from the origin to the unit circle point, you need to go right $\cos \theta$ units and up $\sin \theta$ units.

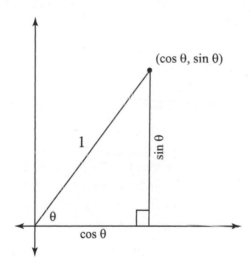

Figure 12.1

The x- and y-coordinates of the point on the unit circle can be treated as the legs of a right triangle whose hypotenuse is 1.

Therefore, $\cos \theta$ and $\sin \theta$ are the lengths of the legs of the right triangle in Figure 12.1 (its hypotenuse has length 1, since it's a radius of the unit circle). According to the Pythagorean Theorem …

$$a^2 + b^2 = c^2$$
$$\left(\cos\theta\right)^2 + \left(\sin\theta\right)^2 = 1^2$$
$$\cos^2\theta + \sin^2\theta = 1$$

All trig functions, not just sine and cosine, are related in some way via the Pythagorean Theorem. This family of three Pythagorean identities is so important that I have given each a name:

- $\cos^2 \theta + \sin^2 \theta = 1$ (The Mama Theorem)
- $1 + \tan^2 \theta = \sec^2 \theta$ (The Papa Theorem)
- $1 + \cot^2 \theta = \csc^2 \theta$ (The Baby Theorem)

Kelley's Cautions

Note that the quantities $(\cos \theta)^2$ and $\cos^2 \theta$ mean exactly the same thing—the second notation is simply a shorter way of writing it that doesn't require parentheses. You shouldn't write $(\cos \theta)^2$ as $\cos \theta^2$ because then it looks like just the θ is squared—that's why you write the exponent next to the function instead.

Keep in mind that I am the only one who uses these names for the Pythagorean identities, so if you ever write "I proved this using the Mama Theorem" without explaining what the heck you're talking about, your math teacher, fellow students, school officials, ancestors, bank tellers, T.G.I. Friday's waitresses, and perhaps mankind at large will think you're nuts (and I won't be around to tell them otherwise.)

Advanced Training: Simplifying Trigonometric Expressions

You now have quite an arsenal of identity-detection tools at your disposal. If a trig expression comes around the corner with a negative sign inside it like sec $(-\theta)$—a mathematical disguise the equivalent of a glued-on fake moustache—you know it's really sec θ behind that terrible makeup job. Sure, secant may look a little different with that negative angle, but you know from basic training that, negative angle inside or not, secant has the same value.

You can even manipulate what you learned in basic training to address new situations. For instance, you could rewrite the expression $1 - \cos^2 \theta$, even though you don't have an identity that contains that *exact* quantity. Notice that the Mama Theorem has both a 1 and a $\cos^2 \theta$ in it. Watch what happens if you subtract $\cos^2 \theta$ from both sides of Mama:

$$\cos^2 \theta + \sin^2 \theta = 1$$

$$\sin^2 \theta = 1 - \cos^2 \theta$$

According to that equation, $1 - \cos^2 \theta$ is actually equal to $\sin^2 \theta$. You should feel free to "massage" the basic identities I've shown you so far, like I did with the Mama Theorem, to generate new information. Just make sure you do the same things to both sides of the identity, or you might make it invalid.

Before you're actually ready to go out and handle identity theft cases, you need just a bit more practice with the identities I showed you. In the following example, I will give you unnecessarily complex trig expressions. Your job will be to cut through the disguise and simplify the expression as much as you can.

Example 2: Simplify each of the following expressions:

a. $\sin(-\theta) \cdot \tan(-\theta) \cdot \cos(-\theta) + \cos^2$

This expression contains two terms added together, but the left term is actually the product of three things, each of which can be rewritten using sign identities:

$$(-\sin\theta)(-\tan\theta)(\cos\theta) + \cos^2\theta$$

Now write everything in terms of the sine and cosine. In other words, write $\tan\theta$ as $\dfrac{\sin\theta}{\cos\theta}$:

$$(-\sin\theta)\left(-\frac{\sin\theta}{\cos\theta}\right)(\cos\theta) + \cos^2\theta$$

Time to multiply the three terms together. The negative signs will disappear since a negative times a negative equals a positive:

$$\left(-\frac{\sin\theta}{1}\right)\left(-\frac{\sin\theta}{\cos\theta}\right)\left(\frac{\cos\theta}{1}\right) + \cos^2\theta$$

$$= \frac{\sin^2\theta \cos\theta}{\cos\theta} + \cos^2\theta$$

Simplify the fraction by canceling $\cos\theta$ in the numerator and denominator.

$$\frac{\sin^2\theta \;\cancel{\cos\theta}}{\cancel{\cos\theta}} + \cos^2\theta = \sin^2\theta + \cos^2\theta$$

Critical Point

By "massaging" the identities, I mean you're allowed to make small changes like you would to any equation, such as adding or multiplying both sides by the same value. Maybe the thought of "massaging Mama" or "massaging Papa" is creepy, but it's often useful, especially for the Pythagorean identities. Plus, Mama's tired shoulders feel better when you're done.

Critical Point

If you ever get stuck simplifying trig expressions or trying to prove trig identities (which you'll do in the next section), try writing everything in terms of sine and cosine or factoring.

You're not quite done yet because $\sin^2 \theta + \cos^2 \theta$ can still be simplified. It's perfectly legal to rewrite it as $\cos^2 \theta + \sin^2 \theta$ (since addition is commutative), which equals 1, according to Mama. Therefore, $\sin(-\theta) \cdot \tan(-\theta) \cdot \cos(-\theta) + \cos^2\theta$ actually equals 1.

b. $\sin^2\left(\dfrac{\pi}{2} - x\right)\left(1 - \sec^2 x\right)$

Right away, the $\left(\dfrac{\pi}{2} - x\right)$ inside the trig function should trigger a red flag in your brain. You need to use a cofunction identity:

$$\cos^2 x(1 - \sec^2 x)$$

Now turn your attention to the $1 - \sec^2 x$ part of the expression. If you massage Papa just a little (his neck muscles are really knotted up) by subtracting 1 from both sides of his identity ($1 + \tan^2 \theta = \sec^2 \theta$), you'll get $\tan^2 \theta = \sec^2 \theta - 1$. Well, $\sec^2 \theta - 1$ is close to what you see in the problem.

If you multiply both sides of the newly massaged Papa by -1, you get $-\tan^2 \theta = -\sec^2 \theta + 1$, or $-\tan^2 \theta = 1 - \sec^2 \theta$. That matches the quantity in the problem exactly, so replace $1 - \sec^2 x$ with $-\tan^2 x$:

$$\cos^2 x(-\tan^2 x)$$

Rewrite in terms of sine and cosine, and simplify:

$$\cos^2 x\left(-\frac{\sin^2 x}{\cos^2 x}\right)$$
$$= -\frac{\cancel{\cos^2 x} \cdot \sin^2 x}{\cancel{\cos^2 x}}$$
$$= -\sin^2 x$$

You've Got Problems

Problem 1: Factor and simplify the expression $\csc^4 \theta - \cot^4 \theta$.

Proving Identities

Finally, your training is over, and you're ready to metaphorically track down credit card imposters by proving trig identities. You'll be given two suspects (expressions separated by an equals sign) who supposedly have the same identity. Your job is to prove that they are indeed equal by manipulating the equations until you get a statement that's true beyond any shadow of a doubt.

There is no one correct approach to proving identities, but there are two popular schools of thought. One approach is what I call the "old school" method—treat the identity like any equation, and solve it by adding, subtracting, and multiplying and dividing the same thing on both sides. Eventually, this turns the equation into a very simple true statement, like "1 = 1" or "$\cos^2 + \sin^2 = 1$." I call this the "old school" approach because there's really nothing new to learn.

Critical Point

Since trig identities are really just equations, you can eliminate all their fractions. Just multiply the identity by its least common denominator.

Another popular approach is the "choose a side" method, in which you take the more complicated-looking side of the equation and manipulate it only, leaving the simpler-looking side alone. Your hope is to turn the complicated side into the simpler one. In Example 2b, you basically proved the identity $\sin^2\left(\dfrac{\pi}{2} - x\right)\left(1 - \sec^2 x\right) = -\sin^2 x$ using the "choose a side" method, manipulating the left side of the equation until it matched the right side. Generally, I prefer the "old school" method (or a mixture of both that leans toward the old school).

Example 3: Prove the following identities:

a. $1 + \tan^2 \theta = \sec^2 \theta$

Even though this is a Pythagorean identity, you can prove that it's true. Rewrite everything in terms of sine and cosine.

$$1 + \frac{\sin^2\theta}{\cos^2\theta} = \frac{1}{\cos^2\theta}$$

Multiply the entire equation by the least common denominator and reduce to eliminate the fractions:

$$\left(\frac{\cos^2\theta}{1}\right)\left[1 + \frac{\sin^2\theta}{\cos^2\theta} = \frac{1}{\cos^2\theta}\right]$$

$$\frac{\cos^2\theta}{1} + \frac{\cancel{\cos^2\theta}\,\sin^2\theta}{\cancel{\cos^2\theta}} = \frac{\cancel{\cos^2\theta}}{\cancel{\cos^2\theta}}$$

$$\cos^2\theta + \sin^2\theta = 1$$

Kelley's Cautions

Even though I proved the Papa identity in Example 3a, it's not absolutely necessary to prove an identity that I've previously told you was true. I proved it only to show you that it could be done (and because it wasn't as obviously true as the Mama Theorem, with its handy right triangle diagram).

The end result is Mama, which (according to basic training) is definitely true, so you're done.

b. $\dfrac{1-\csc\theta}{\cot\theta} = -\dfrac{\cot\theta}{1+\csc\theta}$

Don't you just love eliminating fractions? The thought of multiplying both sides by cot θ(1 + csc θ) and simplifying makes me all tingly inside:

$$\left(\frac{\cot\theta\,(1+\csc\theta)}{1}\right)\left(\frac{1-\csc\theta}{\cot\theta} = -\frac{\cot\theta}{1+\csc\theta}\right)$$

$$\frac{(\cot\theta)\,(1+\csc\theta)(1-\csc\theta)}{\cot\theta} = -\frac{\cot^2\theta\,(1+\csc\theta)}{1+\csc\theta}$$

$$1-\csc\theta+\csc\theta-\csc^2\theta = -\cot^2\theta$$

$$1-\csc^2\theta = -\cot^2\theta$$

Add cot² θ and then csc² θ to both sides of the equation, and you'll get the Baby Theorem:

$$1-\csc^2\theta+\cot^2\theta = 0$$

$$1+\cot^2\theta = \csc^2\theta$$

You've Got Problems

Problem 2: Verify the identity $\cos x - \cos^3 x = \cos^2\left(\dfrac{\pi}{2} - x\right)\cos(-x)$.

Identities with Guts

Occasionally, you'll run across trig expressions with bizarre-looking insides involving addition and subtraction, like sin (x − π), or angles with coefficients, like cos 2θ. Most of the time, trig functions with insides like that should be rewritten as single-angle expressions—trig functions with a simple, single variable inside, like cos x or sin θ.

All of your antifraud training dealt with single-angle expressions, so the process of rewriting the weird expressions allows you to use the identities you already know. Although expressions with strange insides aren't rare, they're much less common than the trig expressions you've worked with so far in this chapter.

Sum and Difference Formulas

Whenever the input of sine or cosine contains a sum or difference, rewrite it using the appropriate formula that follows (appropriately called the sum and difference formulas):

◆ $\sin(\alpha \pm \beta) = \sin\alpha \cos\beta \pm \cos\alpha \sin\beta$

◆ $\cos(\alpha \pm \beta) = \cos\alpha \cos\beta \mp \sin\alpha \sin\beta$

As you may have guessed, the sign $\mp$ means "the opposite of $\pm$," so the two signs in the cosine formula must be opposites, whereas the signs in the sine formula match.

Example 4: Rewrite using a sum and difference formula and simplify: $\cos(\theta - \pi)$.

Solution: Use the cosine difference formula with $\alpha = \theta$ and $\beta = \pi$. Since the original expression contains a $-$, the new sign will be the opposite:

$$\cos(\theta - \pi) = \cos\theta \cos\pi + \sin\theta \sin\pi$$
$$= \cos\theta(-1) + \sin\theta(0)$$
$$= -\cos\theta$$

Critical Point

You can verify the answer to Example 4 by graphing both $y = \cos(\theta - \pi)$ and $y = -\cos\theta$ on the same coordinate axes. You can tell they're equal because they will have the same graphs. In fact, you can use this graphical checking method for any identity.

Double-Angle Identities

If you see double angles (such as 2θ or $2x$) in a trig identity problem, immediately rewrite them in terms of single angles, using a double-angle identity:

◆ $\sin 2\theta = 2\sin\theta \cos\theta$

◆ $\cos 2\theta = \cos^2\theta - \sin^2\theta$

◆ $\cos 2\theta = 2\cos^2\theta - 1$

◆ $\cos 2\theta = 1 - 2\sin^2\theta$

Notice that there's only one sine double-angle formula, but there are three different cosine double-angle formulas to choose from; pick the one containing functions that match those in whatever problem you're working on. (If you feel like it, you can show all three of the cosine formulas are actually equal by "massaging Mama." Just plug $(1 - \sin^2\theta)$ in for $\cos^2\theta$ or $(1 - \cos^2\theta)$ in for $\sin^2\theta$, and prove the identity—it's really easy.)

Example 5: Verify the identity:

$$\frac{\sin 2\theta \cdot \cos 2\theta}{\cos\theta + \sin\theta} = 2\sin\theta\cos^2\theta - 2\sin^2\theta\cos\theta$$

Solution: Rewrite the numerator of the left side using double-angle formulas. Choose the $\cos^2\theta - \sin^2\theta$ version of $\cos 2\theta$ since the rest of the identity has both cosines *and* sines in it. Factor the right side of the equation:

$$\frac{2\sin\theta\cos\theta\left(\cos^2\theta - \sin^2\theta\right)}{\cos\theta + \sin\theta} = 2\sin\theta\cos\theta\left(\cos\theta - \sin\theta\right)$$

Notice that $\cos^2\theta - \sin^2\theta$ is the difference of perfect squares, so factor it and simplify:

$$\frac{2\sin\theta\cos\theta\left(\cos\theta + \sin\theta\right)\left(\cos\theta - \sin\theta\right)}{\cos\theta + \sin\theta} = 2\sin\theta\cos\theta\left(\cos\theta - \sin\theta\right)$$

$$2\sin\theta\cos\theta\left(\cos\theta - \sin\theta\right) = 2\sin\theta\cos\theta\left(\cos\theta - \sin\theta\right)$$

The sides of the equation match, so you've proven the identity.

> **You've Got Problems**
>
> Problem 3: Verify that $\sin 2\theta \cdot \sin\left(\theta + \frac{3\pi}{2}\right) = -2\cos^2\theta\sin\theta$.

The Least You Need to Know

- Basic identities fall into four categories: sign, cofunction, reciprocal, and Pythagorean.

- Trig expressions whose inputs are addition or subtraction problems can be rewritten with sum and difference formulas.

- You should eliminate functions containing double angles using double-angle formulas.

- Although the sine has only one double-angle formula, the cosine has three equivalent (though different-looking) versions.

Chapter 3

Solving Trigonometric Equations

In This Chapter

◆ Evaluating inverse trigonometric functions

◆ Differentiating among exact, specified, and general solutions to trig equations

◆ Solving single- and double-angle equations

◆ Determining solutions to weird trig equations by squaring and substituting identities

Solving an equation evokes a feeling in me unlike any other. No matter how many years I have studied math, and regardless of the thousands upon thousands of problems I've solved, one thought always surfaces just before I square off against an equation: "Will this be the one that stymies me?"

Whether or not you breathe a sigh of relief every time you solve an equation or charge bravely into the fray (a la Mel Gibson in *Braveheart*), equations are unique because once the dust has settled, you can always check your answer to see if you actually got it right. How to celebrate a correct answer is up to you—flamboyantly (pumping your fist like Tiger Woods winning the Masters, or cupping your hand to your ear like wrestling legend Hulk Hogan, as if to admonish the crowd—"Did you see what I just did? I can't hear you cheering!"— or subtly (perhaps with a crooked grin that belies the underlying emotion best described as "I just *destroyed* that problem! I am invincible! You can't stop me. You can only hope to contain me!"), it's time to face off against equations again, this time with trig expressions thrown in.

Inverse Trigonometric Functions

We all know and love inverse functions as the posse that rides down and catches functions we need eliminated. Without exponentiating and e^x (as you learned in Chapter 9), you could never rid the equation $\ln 3x = 5$ of that troublesome natural log, holding the $3x$ captive inside.

Trig equations will be full of other captor expressions, such as sine and cosine, so you'll need inverse trigonometric expressions to cancel them out as well. Here's the only problem: For a function to actually have an inverse, it must pass the horizontal line test, and not a single one of the trigonometric functions do. They're all periodic functions, remember, so their graphs repeat over and over, guaranteeing that not only will each y-value on the graph not be unique, but, in fact, they will repeat an infinite number of times.

Talk the Talk

A **restricted graph** is a small segment of a larger (often periodic) graph that passes the horizontal line test and, therefore, has a valid inverse function.

Fortunately, mathematicians are clever people and found a way to overcome this obstacle using *restricted graphs*. In Figure 13.1, I've highlighted a small section of the sine graph—the interval $\left[-\frac{\pi}{2}, \frac{\pi}{2}\right]$—which is known as the restricted sine graph.

Notice that the restricted sine graph *does* pass the horizontal line test, so as long as you restrict the domain of sine to $\left[-\frac{\pi}{2}, \frac{\pi}{2}\right]$, you can define its inverse function (labeled $\sin^{-1} x$ or $\arcsin x$), which will then have a restricted range of $\left[-\frac{\pi}{2}, \frac{\pi}{2}\right]$. Like all inverse functions, the domain of the trig function becomes the range of its inverse function, and vice versa. Notice that an inverse trig function always outputs an angle value (instead of a number, like a regular trig function).

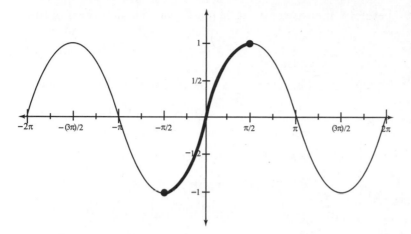

Figure 13.1

Even though y = *sin* x *doesn't pass the horizontal line test, the darkened portion on*

$\left[-\dfrac{\pi}{2}, \dfrac{\pi}{2} \right]$ *does.*

Every trig function can be restricted so that it has an inverse function. Three of them (sine, cosecant, and tangent) use the interval of $\left[-\dfrac{\pi}{2}, \dfrac{\pi}{2} \right]$, illustrated by Figure 13.1. The other three (the cosine, secant, and cotangent) are restricted by the interval $[0,\pi]$. Be sure to remember which interval goes with which function.

Think of the restricted ranges for the inverse trig functions in terms of the quadrants they represent. Functions whose ranges are restricted to $\left[-\dfrac{\pi}{2}, \dfrac{\pi}{2} \right]$ will output angles only in quadrants I and IV; ranges restricted to the interval $[0,\pi]$ output angles in quadrants I and II.

Kelley's Cautions

An inverse trig function can be denoted using either $^{-1}$ behind the function or "arc" in front of it. For instance, the inverse of secant can be written as $\sec^{-1}$ or "arcsec." I prefer the arc method of writing inverses (and will use it exclusively) because too often students misinterpret the $^{-1}$ as a −1 exponent.

Example 1: Evaluate each expression, making sure you give answers in the correct restricted range:

a. $\arccos \dfrac{1}{2}$

This expression asks "What angle has a cosine value of $\dfrac{1}{2}$?" According to the unit circle, cosine has that value at $\theta = \dfrac{\pi}{3}$ and $\theta = \dfrac{5\pi}{3}$.

Although you might be tempted to give both as an answer, remember that $y = \arccos \theta$ is a function, so it can have only one output, not two. Furthermore, you know that arccosine has a restricted range of $[0,\pi]$ (since cosine has a restricted domain of $[0,\pi]$). Therefore, the answer must be $\theta = \frac{\pi}{3}$, since it alone falls in the correct range.

b. $\arctan-\frac{\sqrt{3}}{3}$

This expression asks "What angle has a tangent value of $-\frac{\sqrt{3}}{3}$?", and it's a little tricky. You need to look for unit circle angles whose sine divided by their cosine equals $-\frac{\sqrt{3}}{3}$ (since $\tan\theta = \frac{\sin\theta}{\cos\theta}$).

After some searching, you'll find that $\theta = \frac{5\pi}{6}$ and $\theta = \frac{11\pi}{6}$ are the only angles that fit that description:

$$\tan\frac{5\pi}{6} = \tan\frac{11\pi}{6} = -\frac{\frac{1}{2}}{\frac{\sqrt{3}}{2}} = -\frac{1}{\sqrt{3}} \cdot \frac{\sqrt{3}}{\sqrt{3}} = -\frac{\sqrt{3}}{3}$$

Notice that you have to rationalize the original tangent answer of $-\frac{1}{\sqrt{3}}$ to get $-\frac{\sqrt{3}}{3}$. However, the angles you get are very troublesome; in fact, neither one is a valid output of arctangent!

The answer can't be $\theta = \frac{5\pi}{6}$ because that's in the second quadrant (and the arctangent has a restricted range of quadrants I and IV). To add insult to injury, even though $\theta = \frac{11\pi}{6}$ is in the fourth quadrant, it's definitely too big to fit on the interval $\left[-\frac{\pi}{2}, \frac{\pi}{2}\right]$. To remedy that, you need to find a smaller coterminal angle by subtracting 2π:

$$\frac{11\pi}{6} - 2\pi = \frac{11\pi}{6} - \frac{12\pi}{6} = -\frac{\pi}{6}$$

Therefore, $\arctan-\frac{\sqrt{3}}{3} = -\frac{\pi}{6}$. It might feel weird to give a final answer that's a negative angle, but your hand is forced by those feisty restricted ranges.

You've Got Problems

Problem 1: Evaluate $\operatorname{arc\,csc}-\sqrt{2}$.

Is That Your Final Answer?

Trigonometric equations are a little different than the equations you've dealt with so far. Since they all involve periodic functions, just about every equation you come across will have an infinite number of answers. To deal with the sheer magnitude of answers a trig equation contains, you'll be asked to express your solutions in one of three ways:

- **Exact form.** A trig equation will have exactly one answer for each inverse function you use to solve it; an answer in exact form provides only those restricted range angles, nothing more.

- **Specified form.** When a problem asks you to provide all valid solutions on a specified interval—usually something like $[0,2\pi)$—it wants you to ignore that whole "Hey, buddy, only one answer allowed!" rule and crank out all the solutions you can find between the endpoints you're given.

- **General form.** An answer in general form accounts for the infinite number of coterminal angles that correspond to each of your solutions. Practically, it's all the solutions on the interval $[0,2\pi)$ with $+ 2k\pi$ tacked onto the end of each.

There are other valid, but slightly less common, ways to write solutions to trigonometric equations, such as mime form (in which you don't talk and pretend that you and the solutions are stuck in a big, invisible box) and chloro-form (accidentally inhaling a rag soaked with your answers causes you to momentarily lose consciousness), but I'll stick with the major ones for now.

Example 2: Assume that you've solved a trig equation and end up with the result $\theta = \arctan -1$. Express your solution to the equation in three ways: in exact form, in general form, and on the interval $[0,2\pi)$.

Solution: Start with the specified form of the answer, and consider only unit circle answers. To have a tangent of 1, an angle would have to have equal sine and cosine values (so that the fraction $\frac{\sin\theta}{\cos\theta}$ reduces to 1). To get a tangent of -1, the sine and cosine values of the angle would have to be opposites. Only $\theta = \frac{3\pi}{4}$ and $\theta = \frac{7\pi}{4}$ fit that description, so that's the specific form of the answer: $\theta = \frac{3\pi}{4}, \frac{7\pi}{4}$.

Once you know the specific solution, the other two are simple. The general solution just tacks $+ 2k\pi$ onto the end of each answer: $\theta = \frac{3\pi}{4} + 2k\pi$, $\theta = \frac{7\pi}{4} + 2k\pi$. The exact solution is $\theta = -\frac{\pi}{4}$, the only coterminal angle to $\theta = \frac{7\pi}{4}$ that falls in the interval $\left[-\frac{\pi}{2}, \frac{\pi}{2} \right]$, the restricted range for the arctangent. (Disregard $\theta = \frac{3\pi}{4}$ because it's located in the second quadrant. Remember, exact arctangent answers are always located in quadrants I and IV.)

> ### You've Got Problems
>
> Problem 2: Write the solution(s) to the equation $y = \arccos 1$ in the following ways: (a) on the interval $[0, 2\pi)$; (b) in exact form; and (c) in general form.

Basic Equations

Finally, it's time to actually solve some trigonometric equations. I think you'll be pleasantly surprised to discover that simple trig equations are solved exactly like simple linear equations. The only difference is this: Instead of isolating the x and then dividing both sides by a number to eliminate a coefficient, you'll isolate the θ and take the inverse of the trig function to eliminate it.

Example 3: Find the exact solution for $3\csc\theta + 5 = 11$.

Solution: Isolate the trig term by subtracting 5 from both sides and then dividing everything by 3:

$$3\csc\theta = 6$$
$$\csc\theta = 2$$

> **Kelley's Cautions**
>
> Don't panic if you end up with something that's not on the unit circle. To evaluate something like $\theta = \arccos\frac{1}{3}$, just type it into your graphing calculator to get an exact solution of $\theta = 1.231$. (Make sure you're in radian mode.)

If the cosecant of θ is 2, the sine of θ must be the reciprocal of 2:

$$\sin\theta = \frac{1}{2}$$

Take the inverse sine function (arcsine) of both sides to cancel out sine:

$$\arcsin(\sin\theta) = \arcsin\frac{1}{2}$$
$$\theta = \arcsin\frac{1}{2}$$

Although both $\theta = \dfrac{\pi}{6}$ and $\theta = \dfrac{5\pi}{6}$ have a sine value of $\dfrac{1}{2}$, you're prompted for the exact solution; toss out the second quadrant angle for a final answer of $\theta = \dfrac{\pi}{6}$.

You've Got Problems
Problem 3: Determine the general solution of 2(cot θ + 3) = 6.

Quadratic Equations

You already know three different ways to solve quadratic equations (factoring, using the quadratic formula, and completing the square), and that's plenty—there's no need to learn a new technique just because you'll spot trig functions in there. That would be like learning a new way to walk just because you were wearing a different-color shirt (even though I've always felt better sauntering in a blue shirt, moseying in a white one, and ambling in green). If I were you, I'd pare it even further and use only two strategies: factoring and using the quadratic formula.

The only difference between the quadratic equations you're used to and the ones you're about to face is the final step. Instead of $x =$, you'll end up with a trig function, like $\tan \theta =$. Just apply an inverse trig function to both sides, and you're finished.

Kelley's Cautions

When solving a quadratic equation by factoring or using the quadratic formula, remember to set the equation equal to 0 first.

Example 4: Determine the exact solution to $4\cos^2 \theta + 11\cos \theta + 6 = 0$.

Solution: This is the same equation as $4x^2 + 11x + 6 = 0$, except that, instead of x, you've got cos θ. Either way, the equation is factorable:

$$(4\cos \theta + 3)(\cos\theta + 2) = 0$$

Set each factor equal to 0, and solve:

$$4\cos\theta + 3 = 0$$
$$\theta = \arccos\left(-\dfrac{3}{4}\right) \quad \text{or} \quad \cos\theta + 2 = 0$$
$$\theta = \arccos(-2)$$

Kelley's Cautions

Because valid arcsine and arccosine inputs must fall on the interval [−1,1], arcsecant and arctangent inputs must belong either to (−∞,−1] or [1,∞). Any values work for tangent and co-tangent, so there's no need to check those.

One of the solutions is invalid. No angle can have a cosine of −2 because the range of cosine is [−1,1]. Think about the untransformed graphs of sine and cosine from Chapter 11; they go from a minimum height of −1 to a maximum height of 1, so a cosine value of 2 doesn't make sense. Therefore, discard $\theta = \arccos(-2)$.

The other solution, $\theta = \arccos\left(-\frac{3}{4}\right)$, definitely falls in the interval [−1,1], but it doesn't match any of the unit circle values you've memorized. Type it into your calculator to get the answer in exact form: $\theta = 2.419$.

You've Got Problems

Problem 4: List all of the solutions to $2\sin^3 \theta - \sin^2 \theta = \sin \theta$ on the interval $[0, 2\pi]$.

Mismatched and Needy Equations

So far in this chapter, every equation you've solved has contained only one function (like $\sin \theta$ or $\cos \theta$), but the upcoming equations will contain both at once! I classify any equation containing nonmatching functions as "needy" because (like a high-maintenance mate or a friend with an entire American Tourister set of emotional baggage), they need more attention than normal and must be handled with kid gloves. Thankfully, the Pythagorean identities come in very handy.

Needy Equations with Squares

If one or more of the unmatched functions in an equation is squared and both belong to the same Pythagorean identity (like the tangent and secant both belong to Papa), use that identity to put the entire equation in terms of a single function.

Kelley's Cautions

Unless you're specifically told otherwise, assume that any angles in an equation are measured in radians.

Example 5: Find the exact solution for $5\cos \theta = \sin^2 \theta - 4$.

Solutions: This equation contains both the sine and cosine (different functions that both belong to the same Pythagorean identity, Mama), and the sine function is squared. If you massage Mama just a little bit (by subtracting $\cos^2 \theta$ from both of her sides), you'll end up with $\sin^2 \theta = 1 - \cos^2 \theta$. This means the $\sin^2 \theta$ in the equation can be replaced with $1 - \cos^2 \theta$:

$$5\cos\theta = \sin^2\theta - 4$$
$$5\cos\theta = \left(1 - \cos^2\theta\right) - 4$$
$$5\cos\theta = -\cos^2\theta - 3$$

Excellent—the entire equation is now in terms of the cosine and is no longer needy. You're left with a quadratic equation, so set it equal to 0:

$$\cos^2\theta + 5\cos\theta + 3 = 0$$

Unfortunately, that sucker's not factorable, so you'll need to use the quadratic formula; it'll look the same as in the quadratic equations of old, except that in place of $x =$, you'll write the function ($\cos\theta =$) and apply the arccosine to finish:

$$\cos\theta = \frac{-5 \pm \sqrt{5^2 - 4(3)}}{2}$$
$$\cos\theta = \frac{-5 \pm \sqrt{13}}{2}$$
$$\theta = \arccos\left(\frac{-5 - \sqrt{13}}{2}\right), \arccos\left(\frac{-5 + \sqrt{13}}{2}\right)$$

The first solution is invalid since the input is outside the interval $[-1,1]$: $\frac{-5 - \sqrt{13}}{2} \approx -4.303$. However, the second value is fine, so use your calculator to evaluate it:

$$\arccos\left(\frac{-5 + \sqrt{13}}{2}\right) \approx \arccos\left(-0.697224362268\right) \approx 2.342$$

You've Got Problems

Problem 5: Find all solutions on the interval $[0, 2\pi)$ for the equation $2\tan^2\theta = \sec^2\theta$.

Needy Equations Without Squares

If an equation has unmatched functions from the same Pythagorean identity but no squares, introduce them yourself by squaring both sides of the equation and applying the corresponding identity.

Kelley's Cautions

When you square both sides of an equation, there's a very good chance that you're introducing false solutions, so make sure you check your answers once you're finished and toss out any false ones.

Example 6: Find the general solution of csc θ = 1 + cot θ.

Solution: Both csc θ and cot θ are part of Baby, but neither is squared, so Baby is kept at bay until you square both sides. Keep in mind that $(1 + \cot \theta)^2$ does *not* equal $1 + \cot^2 \theta$, just like $(a + b)^2$ does *not* equal $a^2 + b^2$, but instead equals $a^2 + 2ab + b^2$:

$$(\csc\theta)^2 = (1+\cot\theta)^2$$
$$\csc^2\theta = 1+2\cot\theta + \cot^2\theta$$

Notice that the right side of the equation contains $1 + \cot^2 \theta$, which (according to Baby) equals $\csc^2 \theta$:

$$\csc^2 \theta = \csc^2 \theta + 2\cot \theta$$

Subtract $\csc^2 \theta$ from both sides, and solve:

$$0 = 2\cot\theta$$
$$0 = \cot\theta$$
$$\theta = \text{arccot}\,0$$

Cotangent equals 0 whenever its numerator equals 0: $\theta = \dfrac{\pi}{2}, \dfrac{3\pi}{2}$. However, if you plug the solutions back into the original equation, csc θ = 1 + cot θ, only $\theta = \dfrac{\pi}{2}$ works. (Substituting $\dfrac{3\pi}{2}$ for θ gives you –1 = 1.) Therefore, the final answer in general form is $\theta = \dfrac{\pi}{2} + 2k\pi$.

You've Got Problems

Problem 6: Give the exact solution to the equation cos θ + sin θ = 1. (Hint: Move one of the functions to the right side of the equation before squaring.)

Equations with Multiple Angles

If one or more trig expressions in an equation contain multiple angles (like 2θ rather than plain old θ), it barely affects the solution. In fact, the strange angle comes into play only *after* you've used an inverse trig function at the very end of the problem.

Watch how minor the adjustments are as I work through the following example. I'll tell you when the multiple angle has to be dealt with and show you what to do about it.

Example 7: Give all solutions to the equation $2\cos^2 3\theta - 1 = 0$ on the interval $[0, 2\pi)$.

Solution: Forget about the 3θ for now; pretend that it's just a θ in there, if that helps. Isolate the trig function by adding 1 to both sides of the equation and dividing everything by 2:

$$\cos^2 3\theta = \frac{1}{2}$$

Now take the square root of both sides; if you rationalize the right side when you're done, it will better resemble a familiar unit circle value:

$$\sqrt{\cos^2 3\theta} = \pm\sqrt{\frac{1}{2}}$$

$$\cos 3\theta = \pm\frac{1}{\sqrt{2}}$$

$$\arccos(\cos 3\theta) = \arccos\left(\pm\frac{\sqrt{2}}{2}\right)$$

$$3\theta = \frac{\pi}{4}, \frac{3\pi}{4}, \frac{5\pi}{4}, \frac{7\pi}{4}$$

Now it's time to deal with that pesky multiple angle 3θ. Since the coefficient is 3, you should write *three times as many answers* as you regularly would. So far, there are four answers there (each of the $\frac{\pi}{4}$ angles on the unit circle has a cosine of either $\frac{\sqrt{2}}{2}$ or $-\frac{\sqrt{2}}{2}$), so you should end up with $4 \cdot 3 = 12$ total answers. To accomplish this, add 2π to each of your solutions two separate times:

$$3\theta = \frac{\pi}{4}, \frac{3\pi}{4}, \frac{5\pi}{4}, \frac{7\pi}{4}, \frac{9\pi}{4}, \frac{11\pi}{4}, \frac{13\pi}{4}, \frac{15\pi}{4}, \frac{17\pi}{4}, \frac{19\pi}{4}, \frac{21\pi}{4}, \frac{23\pi}{4}$$

Whoa! That's a big heaping serving of angles. In case you're still wondering where they came from, here's what I did. To the first angle $\left(\frac{\pi}{4}\right)$, I added 2π twice: $\frac{\pi}{4} + \frac{8\pi}{4} = \frac{9\pi}{4}$ and $\frac{9\pi}{4} + \frac{8\pi}{4} = \frac{17\pi}{4}$. Then I did the same thing with the other three original solutions $\left(\frac{3\pi}{4}, \frac{5\pi}{4}, \text{and } \frac{7\pi}{4}\right)$.

All you have to do to finish this problem is to solve for θ by multiplying every single thing in that conga line of fractions by $\frac{1}{3}$, to eliminate the coefficient of θ:

$$\theta = \frac{\pi}{12}, \frac{3\pi}{12}, \frac{5\pi}{12}, \frac{7\pi}{12}, \frac{9\pi}{12}, \frac{11\pi}{12}, \frac{13\pi}{12}, \frac{15\pi}{12}, \frac{17\pi}{12}, \frac{19\pi}{12}, \frac{21\pi}{12}, \frac{23\pi}{12}$$

You've Got Problems
Problem 7: Give the general solution to $2\cos 2\theta + 1 = 0$.

The Least You Need to Know

♦ The restricted range of the arcsine, arctangent, and arccosecant functions is $\left[-\dfrac{\pi}{2}, \dfrac{\pi}{2}\right]$; the restricted range for arccosine, arccotangent, and arccosecant is $[0, \pi]$.

♦ A trig equation has an exact, specified, and general solution.

♦ You can use Pythagorean identities to get an entire equation in terms of one trig function.

♦ The coefficient of a multiple angle tells you how many coterminal angles should be included as solutions to the equation.

14

Oblique Triangle Theorems

In This Chapter

- Assigning reference angles to oblique angles
- Evaluating trigonometric functions of oblique angles
- Calculating side and angle measurements for oblique triangles
- Determining the area of an oblique triangle

If you don't know what the word *oblique* means, the title of this chapter and every one of the preceding bullets felt like a punch to the face (or maybe like actual bullets). No need to worry—the term *oblique* simply means that a right angle is not involved. Therefore, the only thing you know about an oblique angle is that it definitely does not measure 90°, and an oblique triangle does not contain a right angle.

Did you notice that I tossed a degree measurement in there? Feels like old times when you see an angle without a π sticking out of it, and a few of the topics in this chapter are going to go retro, involving degrees more often than radians. The reason is this: We're going to give the triangles a little bit more freedom. That means you'll see a lot of angles that aren't in standard position, and even a bunch of angles that are nowhere near the coordinate plane at all.

Talk the Talk

The term **oblique** describes something that does not involve a right angle. For instance, an oblique angle does not measure 90°, and an oblique triangle cannot be a right triangle. It's also what you'd say to your pet monkey, Bleek, if he did something bad, as in "Oh, Bleek, why did you give me an incurable jungle virus that causes me to hallucinate and sweat blood?"

When I was in middle school, every girl I knew had a white hooded sweatshirt with an airbrushed picture on the back. It usually featured a beach scene with seagulls, silhouettes of people holding hands, and this (supposedly) philosophical thought written in giant, flowing script: "If you love something, set it free. If it comes back, then it's yours, but if it doesn't, it never was."

You now know enough about trig to set angles free of standard position, just as the shirt so wisely advised. As the theme to TV's *Love Boat* taught you, "Let it float, it floats back to you." Get ready for all sorts of "exciting and new" things, such as applying trig functions to triangles that don't have a right angle in them and calculating areas that were impossible back in geometry class.

Checking References for Unfriendly Angles

Until now, everything I've showed you about trig has, in some way, involved right angles. From the very definitions of the sine and cosine (as ratios of SOHCAHTOA) to the values from the unit circle, every time you turned around, you saw a right triangle. However, since trigonometry means "the measurement of triangles," not "the measurement of only right triangles," you had to figure this day was coming.

Talk the Talk

A **reference angle** is an acute angle whose vertex lies on the origin, with one of its sides on either the positive or negative *x*-axis. It's part of a right triangle whose trig values match those of an angle in an oblique triangle.

To make the leap of faith to oblique angles and triangles, you'll use something called a *reference angle*—an angle in a right triangle that will have the same trig values as an oblique angle—so, actually, not a whole lot of faith is involved.

Behold the Bowtie

Given an angle whose terminal side lies in any quadrant, there is a unique right triangle that generates the necessary reference angle for the angle. When you draw all four triangles at the same time, you get something that looks like a bowtie (as you can see in Figure 14.1). The actual reference angles you'll use are the acute angles I've numbered.

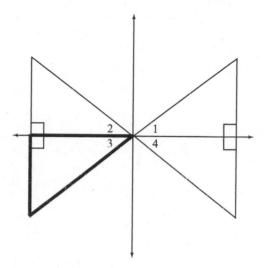

Figure 14.1

The bowtie is made up of four right triangles, whose reference angles are 1, 2, 3, and 4. However, you'll use just one triangle at a time; for instance, any angle with a terminal side in the third quadrant would use the darkened triangle with reference angle 3.

To find a reference angle for an oblique angle, follow these steps:

1. Put the oblique angle in standard position (if it isn't already).

2. Draw a vertical segment connecting the terminal side to the x-axis, forming one quarter of the bowtie from Figure 14.1.

3. The acute angle (angle 1, 2, 3, or 4 in Figure 14.1) formed by the terminal side and the x-axis will have the same trig values as the original angle and is, therefore, the reference angle.

This is how big angles (too big to fit into a right triangle) still manage to have trig values for sine, cosine, tangent, and the rest. Their smaller, acute friends (who belong to valid right triangles) vouch for them, saying things like, "Hey, you'll like this guy—he's got the same work ethic, morals, and cosine values as I do. If you're not sure whether to hire him, I'll act as his reference and tell you that there's nothing to worry about."

Example 1: Determine the measure of the reference angle for each of the following angles:

a. $\alpha = 139°$

Sketch α in standard position, and draw a vertical segment from it down to the x-axis, as demonstrated in Figure 14.2. This forms the second quadrant piece of the bowtie whose acute angle, β, is the reference angle for α.

Figure 14.2

In second-quadrant angles such as α, the reference angle (β, in this diagram) is also the supplement.

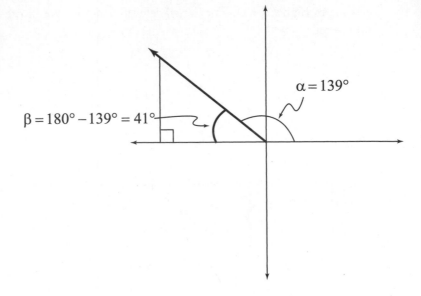

$$\beta = 180° - 139° = 41°$$

α = 139°

It's easy to calculate β, since α + β = 180°. Therefore, 180° − 139° = 41°. Don't ask yourself "Is this angle traveling clockwise or counterclockwise?" because it doesn't matter—a reference angle always has a positive sign.

If the trig function values for an angle and its reference angle don't match exactly, they will differ only by a negative sign (meaning the values are either equal or opposites). We'll deal with that in the next section.

How'd You Do That?

Here are the shortcuts for calculating a reference angle (β) for an angle θ without having to draw a picture each time. I've used radians, but the same process works for degrees—just replace π with 180°:

- Quadrant I: No reference angle required
- Quadrant II: β = π − θ
- Quadrant III: β = θ − π
- Quadrant IV: β = 2π − θ

These work only if 0 ≤ θ < 2π. Therefore, you'll need to calculate a coterminal angle for θ if it doesn't fall within those boundaries.

b. $\theta = \dfrac{9\pi}{7}$

Where do you draw the terminal side for θ? Divide 9 by 7 to get a more descriptive coefficient: $\dfrac{9\pi}{7} \approx 1.286\pi$. Since $\pi = 1.0\pi$ and $\dfrac{3\pi}{2} = 1.5\pi$ (the boundaries for the third quadrant), 1.286π must fall somewhere in there. Draw the angle in standard position (don't worry about drawing it exactly—as long as you're in the right quadrant, that's all that matters), and add the vertical line connecting its terminal side to the x-axis, like I've done in Figure 14.3.

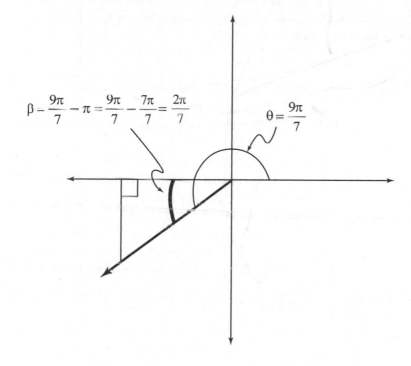

Figure 14.3

The reference angle (β) is the piece of θ that lies in the third quadrant.

$\beta - \dfrac{9\pi}{7} - \pi = \dfrac{9\pi}{7} - \dfrac{7\pi}{7} = \dfrac{2\pi}{7}$

$\theta = \dfrac{9\pi}{7}$

The right triangle pictured in Figure 14.3 is the third-quadrant piece of the bowtie, and the reference angle, β, is marked with a dark arc. Unlike the previous problem, this time the reference angle is actually a small part of θ—the part that extends into the third quadrant. To calculate β, simply subtract π from θ; this removes all of the angle that's contained in quadrants I and II, leaving only the reference angle $\beta = \dfrac{2\pi}{7}$.

You've Got Problems

Problem 1: Determine the reference angle for $\theta = \frac{13\pi}{7}$.

Tying in Trig Functions

Once you've mastered the art of reference angles and wearing a bowtie, you can calculate tons more trig function values than you ever could before. You can even conquer trig functions that are not on the unit circle, at least in some select cases. Check out the following example to see what I mean.

Example 2: If $\tan\theta = \frac{2}{5}$ and $\sin\theta < 0$, evaluate $\cos\theta$.

Critical Point _____

Back in Chapter 11, I showed you a coordinate plane whose quadrants were labeled with the trig functions that were positive there (Figure 11.10). Use that diagram to help you determine the quadrants for these types of problems.

Solution: This problem is pretty easy if you use reference angles and triangles. In fact, you don't even have to figure out what θ actually is! You're given the tangent value of the angle (notice that it's positive—that's very important) and told that the sine value of the angle is negative. Based on this information, you need to decide which quadrant contains the angle. Tangent is positive only in quadrants I and III, and sine is negative in quadrants III and IV. Therefore, θ must be in quadrant III, so draw a third-quadrant bowtie right triangle with reference angle θ (illustrated in Figure 14.4).

Figure 14.4

Because you go left and down to travel from the origin to the end of the hypotenuse, both the horizontal and vertical legs of the triangle should be labeled as negative numbers.

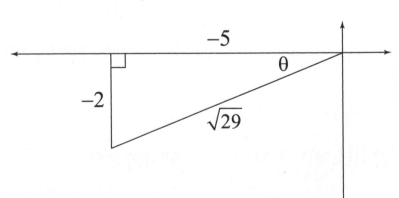

Since you know that $\tan\theta = \frac{2}{5}$ (and that tangent is equal to the opposite side divided by the adjacent side), the side of the triangle opposite θ must have a length of 2, and the adjacent side must have a length of 5. However, since you're using a coordinate plane, the direction in which those legs travel from the origin is extremely important. Whenever you travel left or down, those directions must be indicated by a negative number; in this problem, you travel both left and down, so both numbers are negative.

> **CAUTION**
> **Kelly's Cautions**
>
> The hypotenuse of a reference triangle will *never* be negative. The only negative sides in these reference triangles will be the horizontal side (if the triangle is in quadrants II or III) and the vertical side (if the triangle is in quadrants III or IV).

Since you've drawn a right triangle, use the Pythagorean Theorem to calculate the missing side, which (like all hypotenuses) must be positive:

$$\left(-2\right)^2 + \left(-5\right)^2 = c^2$$
$$4 + 25 = c^2$$
$$c = \sqrt{29}$$

Once you've got all three sides of the triangle labeled, you're finally ready to answer the actual question posed by the problem: What is $\cos\theta$? Remember, cosine is equal to the adjacent side divided by the hypotenuse, so plug in the values from Figure 14.4:

$$\cos\theta = \frac{\text{adjacentside}}{\text{hypotenuse}}$$

$$\cos\theta = -\frac{5}{\sqrt{29}} \text{ or } -\frac{5\sqrt{29}}{29}$$

You've Got Problems

Problem 2: If $\sec\alpha = \frac{\sqrt{89}}{8}$ and $\csc\alpha < 0$, evaluate $\sin\alpha$.

Obliterating Oblique Triangles

I remember a career counselor in college once telling me, "You should look for a job far, far away from other people because we all know how horrible all math majors' social skills are." I wanted to argue and stand up for my mathematical brothers and sisters, my comrades of the complex numbers, friends in fractions, pals in proof, compatriots swaddled in shirts so wrinkled that they looked like a topographical map of the Pyrenees. But then I realized something—he's right. Most math people are introverted and are rarely dynamic folks.

Critical Point _____

How would you describe one and a half hours? Most people (after wondering why the heck I was even asking such a bizarre question) would probably answer either "1.5 hours" or "1 hour and 30 minutes." Hours are funny things because they work on a base 60 system, meaning that each unit is made up of 60 smaller units (60 minutes make up an hour, and 60 seconds make up a minute).

Angles measured in degrees use the same base 60 system, in which minutes are indicated with a prime symbol (') and seconds are indicated by a double prime ("). In other words, 1.5° (1 and ½ degrees) is equal to 1° 30' (1 degree, 30 minutes).

You won't encounter angles written in degrees and seconds very often, but if you do, it's easy to rewrite them as decimals. Just divide the minutes by 60, divide the seconds by 3,600 (60²), and add them to the whole degrees.

For example, 47° 52' 17" (47 degrees, 52 minutes, 17 seconds) has a decimal equivalent of this:

$$47°52'17'' = 47 + \frac{52}{60} + \frac{17}{3600}$$
$$\approx 47.871°$$

Even though radians use fractions (and they can get ugly), they're better than an unwieldy base 60 system that wrestles with the decimal system. This is a major reason most mathematicians prefer radians to degrees, and it's why I won't include degree measurements in minutes and seconds in this book.

But there's one thing you can't argue about: Math people are dang clever. With a swift strike of reasoning and the birth of reference angles, they've opened a huge door for you. Now you can rest assured that any angle, not just an acute angle trapped inside a right triangle, has a sine and a cosine value (as well as values for the other trig functions).

Although it may seem insignificant, it actually means you're free to leave the coordinate plane and the stifling restrictions of standard position. It's time to deal with free-roaming triangles that aren't glued down to a pair of axes. To top it off, you'll be working with oblique triangles, so all of the handy tricks you learned about calculating the sides and angles of right triangles in Chapter 10 get chucked right out the window. Luckily, two brand-new, shiny theorems will leap in to take their place: the Law of Cosines and the Law of Sines.

Most of the triangles you'll encounter with these new theorems will be labeled a particular way—the angles will be labeled using capital letters (*A*, *B*, and *C*), and the sides will be labeled with lowercase letters (*a*, *b*, *c*). It's not the capitalization that's odd, though; it's the repetition of letters that could be confusing. Why would you use *A* and *a* in the same problem? Actually, matching letters are used to indicate something important: The lowercase letter represents a side *opposite* the uppercase angle. For instance, side *b* is opposite angle *B* in the triangle, and the same goes for the other two pairs of letters, resulting in a triangle that looks like Figure 14.5.

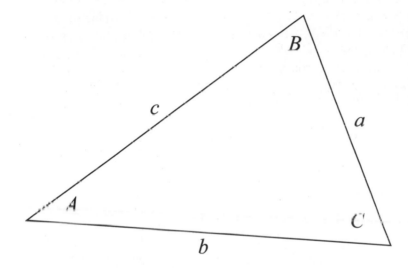

Figure 14.5

In Law of Cosines and Law of Sines problems, sides and angles opposite one another are often labeled with the same letter; the only difference is that sides are lowercase and angles are capitalized.

One of the first decisions you'll need to make in oblique triangle problems is whether to use the Law of Cosines or the Law of Sines; it depends upon what kind of information you're given in the problem. If you know the lengths of all three sides (abbreviated SSS, for side-side-side) or two of the sides and the angle included between them (SAS, short for side-angle-side), you use the Law of Cosines.

The Law of Cosines

The Law of Cosines is a formula specifically engineered for each side of an oblique triangle. Therefore, there are three versions of the Law of Cosines, one for each of its three sides:

- ◆ $a^2 = b^2 + c^2 - 2bc \cos A$

- ◆ $b^2 = a^2 + c^2 - 2ac \cos B$

- ◆ $c^2 = a^2 + b^2 - 2ab \cos C$

The differences in the formulas are minor because each essentially says the same thing: The square of one side is equal to the squares of the other two sides added together minus what you get when you multiply 2 times those other two sides times the cosine of the angle opposite the side you started with.

Example 3: Calculate the largest angle in the triangle with sides of length 5, 8, and 10; write your answer in degrees.

Solution: You're given the length of each side (SSS), so the Law of Cosines should be used. It doesn't matter which is a, b, or c, so you might as well just assign them in order: $a = 5$, $b = 8$, and $c = 10$. You're asked to find the largest angle; remember from geometry that the largest angle and the largest side of a triangle are always opposite one another. Therefore, you're trying to calculate angle C (since $c = 10$ is the largest side). The third version of the Law of Cosines is the only one that contains C, so use it; plug a, b, and c into the formula:

$$c^2 = a^2 + b^2 - 2ab\cos C$$
$$10^2 = 5^2 + 8^2 - 2(5)(8)\cos C$$
$$100 = 25 + 64 - 80\cos C$$
$$100 = 89 - 80\cos C$$
$$11 = -80\cos C$$

Critical Point

Because oblique triangles are not usually drawn on the coordinate plane, the Law of Cosines (and the Law of Sines) problems are usually written in degrees instead of radians, but the formulas will work for either type of angle.

This is now just a basic trig equation. Divide both sides by –80, and take the arccosine to solve for C. Remember to put your calculator in degree mode, as indicated by the original problem.

$$-\frac{11}{80} = \cos C$$
$$-0.1375 = \cos C$$
$$\arccos(-0.1375) = \arccos(\cos C)$$
$$97.903° \approx C$$

How'd You Do That?

The sum of all three angles in any triangle must equal 180°, so no angle inside a triangle can be 180° or larger. Therefore, each angle is either acute (if less than 90°) or obtuse (between 90° and 180°).

If an acute angle is in standard position, its terminal side lands in quadrant I, but if it is obtuse, it lands in quadrant II. Notice that the sign of cosine is different within those quadrants.

In Example 3, $\cos C = -\dfrac{11}{80}$. Since the cosine of C is negative, its terminal side must land in quadrant II; you actually know that C is going to be obtuse before you calculate it. Here's the bottom line: Cosine can tell the difference between acute and obtuse angles because their cosine values have different signs. This is extremely important, because sine cannot tell such angles apart (the sine is positive in both quadrants I and II). This makes the Law of Sines much less useful than the Law of Cosines.

You've Got Problems

Problem 3: Determine the length of the missing side in the following diagram:

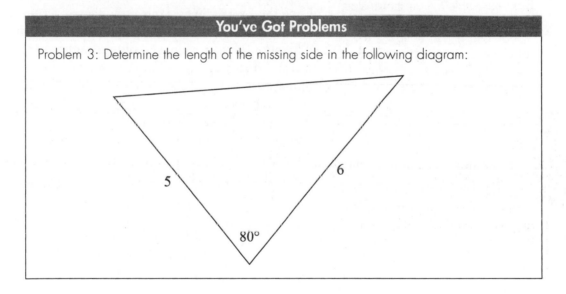

The Law of Sines

Whereas the Law of Cosines specializes in SSS and SAS triangles, the Law of Sines has its own niche: AAS triangles, in which you're given two angle measurements and the length of a side not between those angles. You can also use it when you're given an SSA triangle (textbooks usually write the A last instead of first in this abbreviation, for fear of a PG-13 rating), where you're given two sides and an angle not formed by those sides. However, the Law of Sines won't always work in those cases, so apply it only if you have no other choice.

Happily, the Law of Sines is much easier to remember because its formula is just three fractions set equal to one another:

$$\frac{\sin A}{a} = \frac{\sin B}{b} = \frac{\sin C}{c}$$

In other words, the sine of any angle divided by its opposite side is equal to the sine of any other side divided by its opposite side. Some textbooks define the Law of Sines as $\frac{a}{\sin A} = \frac{b}{\sin B} = \frac{c}{\sin C}$ but that's really the same thing. You're allowed to take the reciprocal of an equation's sides without screwing anything up.

Critical Point _____

If you're asked to measure multiple parts of a triangle in the same problem, you should start by trying to calculate the largest side and angle, using the Law of Cosines, if necessary. Once that's done, finish the problem using the simpler and shorter Law of Sines.

There's only one thing you must keep in mind when using the Law of Sines. Unlike the Law of Cosines, when you use the Law of Sines, you cannot tell the difference between acute and obtuse angles, and you will incorrectly identify obtuse angles. So, the moral of the story is this: If you're trying to identify an angle's measure and there's *any possibility at all* that the angle is obtuse, you must calculate it using the Law of Cosines instead of the Law of Sines.

Example 4: Determine the measures of *A*, *C*, and *c* in the following diagram. Assume that *A* is an acute angle, so there's no fear about needing to use the Law of Cosines.

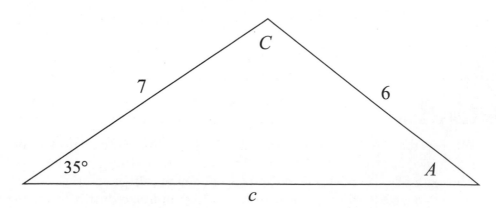

Solution: There's a lot to do here, so start by identifying the measurements you know. You are given only one angle measurement; angles A and C are already labeled in the triangle, so the known angle must be angle B: $B = 35°$. You also know the lengths of two sides: $a = 7$ and $b = 6$. So, you've got two sides and an angle that's not formed by those two sides, an SSA triangle.

Uh oh, I warned you not to use the Law of Sines on SSA triangles unless you had no other choice. Unfortunately, we don't have AAS (which could be solved with the Law of Sines easily) or either Law of Cosines form (SSS or SAS). Therefore, you're forced to use the Law of Sines. Don't worry—it'll all work out in the end.

To use the Law of Sines, identify one angle/side pair whose measurements you know (in this case, you should use $b = 6$ and $B = 35°$). Then pick out an angle/side pair that you know *something* about. In other words, choose $a = 7$ and A because at least you know the length of the side. (Right now, we're completely in the dark about c and C, but that will change.) Plug a, b, sin A, and sin B into the Law of Sines.

Critical Point

Even though the Law of Sines is defined as three fractions set equal to one another, you'll use only two fractions at a time. Just choose any two of the three, and set them equal to one another.

$$\frac{\sin 35°}{6} = \frac{\sin A°}{7}$$

Multiply both sides of the equation by 42 to eliminate the fractions, and then solve for A:

$$7 \sin 35° = 6 \sin A°$$

$$\frac{7(0.573576436)}{6} = \sin A$$

$$0.669172509 = \sin A$$

$$A = \arcsin(0.669172509) \approx 42.003230743°$$

Don't round A yet! There's lots more work to do, and you shouldn't round until the end, to ensure the accuracy of the other measurements to come. Now that you know A and B, you can easily figure out C, since the sum of the angles of a triangle must add up to 180°:

$$35° + 42.003230743° + C = 180°$$

$$C = 102.996769257°$$

All that's left to calculate is side c, so use the Law of Sines again to figure it out:

$$\frac{\sin 35°}{6} = \frac{\sin 102.996769257°}{c}$$

$$c \cdot \sin 35° = 6\left(\sin 102.996769257°\right)$$

$$c = \frac{5.84629648544}{\sin 35°}$$

$$c \approx 10.192706874$$

Now that all the missing measurements are present and accounted for, feel free to round your answers: $A = 42.003°$, $C = 102.997°$, and $c = 10.193$.

You've Got Problems

Problem 4: Determine the measures of all three angles in the following diagram:

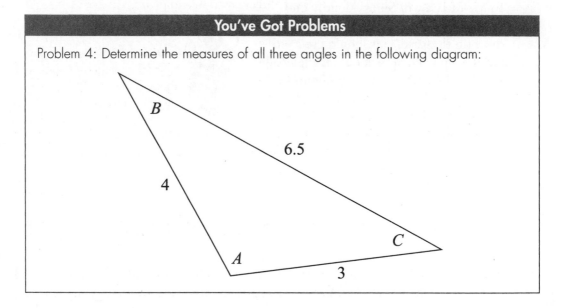

Calculating an Oblique Triangle Area

If I sold you a mystical medallion that protected you from any blue-eyed werewolf that tried to attack (and, for some reason, you feared such an attack was imminent), you'd think that medallion was pretty nifty until you found out that 95 percent of werewolves had brown eyes. A similar thing happens all too often in math. You're given math formulas that work only in very specific situations—and no matter how handy they are in those situations, it doesn't make up for how totally and hopelessly useless they are in every other circumstance.

The trig ratios (sine, cosine, and tangent) worked great when you were calculating the sides and angles of right triangles, but let's be honest: The majority of triangles aren't right triangles. So aim the magical SOHCAHTOA medallion at oblique triangles all you like, and you're still going to get eaten alive by the first oblique triangle that jumps out at you from behind the bushes. Once you learn the Laws of Cosines and Sines, however, you can calculate the pieces of any triangle and walk safely at night, even during a full moon.

The age-old formula for the area of a triangle is $A = \frac{1}{2}bh$, one half of the triangle's base times its height. Again, the formula is fundamentally flawed: The base and the height of the triangle have to be perpendicular to one another. Unless you're given a right triangle—or at least a triangle with a height that's easy to calculate—you're sunk. Until now, that is. You're about to learn two more techniques to calculate triangle area: one that works if you know the lengths of two sides and the angle they form (SAS), and one that works if you know the lengths of all three sides (SSS).

Critical Point

Both new area formulas work whether you're measuring angles in degrees or radians—no adjustments are necessary.

SAS Triangles

If a triangle has sides a and b that form an angle C, the area of that triangle is this:

$$A = \frac{1}{2}ab \cdot \sin C$$

It looks sort of like the old "one-half base times height" formula, except that a and b don't have to be specific sides of the triangle. All they have to do is form angle C.

Example 5: Determine the area of the following triangle:

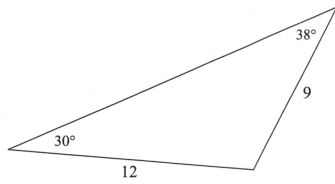

Solution: To apply the new formula $A = \frac{1}{2}ab \cdot \sin C$, you need to know the measurement of the angle formed by the sides whose lengths are 9 and 12. The angle you plug into the formula has to be formed by the sides you plug in. Subtract both of the other angles in the triangle from 180° to find out what the remaining angle is: 180° − 38° − 30° = 112°.

Now plug everything into the formula:

$$\text{Area} = \frac{1}{2}ab \cdot \sin C$$

$$= \frac{1}{2}(9)(12) \cdot \sin 112°$$

$$\approx 54(0.92718385)$$

$$\approx 50.068$$

You've Got Problems

Problem 5: Two sides of a triangle have lengths 4 and 7, and the angle they form measures 40°. What is the area of the triangle?

SSS Triangles

Even if you're not given a single angle measurement in a triangle, you can calculate its area if you know the lengths of all three sides. According to *Heron's Area Formula*, a triangle with sides a, b, and c has area $= \sqrt{s(s-a)(s-b)(s-c)}$ if $s = \frac{a+b+c}{2}$.

Example 6: Triangle ABC has sides that measure 5 cm, 9 cm, and 12 cm; calculate its area, and label your answer using the appropriate unit of measure.

Solution: Start by calculating s:

$$s = \frac{a+b+c}{2} = \frac{5+9+12}{2} = \frac{26}{2} = 13$$

Since there are no opposite angles to worry about, it really doesn't matter which side you label a, b, or c; you'll get the same answer from Heron's Area Formula:

$$\text{area} = \sqrt{s(s-a)(s-b)(s-c)}$$

$$= \sqrt{13(13-5)(13-9)(13-12)}$$

$$= \sqrt{13(8)(4)(1)}$$

$$= 4\sqrt{26} \text{ cm}^2 \approx 20.396 \text{ cm}^2$$

Talk the Talk

Heron's Area Formula states that an oblique triangle with sides a, b, and c has area $\sqrt{s(s-a)(s-b)(s-c)}$ if $s = \frac{a+b+c}{2}$.

You should label your answer as square centimeters (cm²) because the correct area unit of measure is the square of the unit used to measure the sides of the figure.

You've Got Problems

Problem 6: Calculate the area of a triangle whose sides are 9, 13, and 16 inches long.

The Least You Need to Know

◆ Reference angles are acute angles from right triangles that have the same trigonometric values (but possibly different signs) as oblique angles.

◆ There are four different reference triangles, each with its own reference angle and each corresponding to one quadrant of the coordinate plane.

◆ The Laws of Cosines and Sines are used to calculate the angles and sides of oblique triangles.

◆ Only the Law of Cosines can accurately determine the measurement of an obtuse angle.

◆ You can easily calculate the areas of oblique SAS and SSS triangles.

Part 4

Conic Sections

In this part, you'll take one last romp through the land of shapes you knew about long ago (and some that were never mentioned in kindergarten, unless you went to one of those fancy ones where they do yoga and interpretive dance). However, as nostalgic as working with things like circles and ovals may be, precalculus is (of course) going to step in with tons of variables and formulas to ruin the fun.

15

Parabolas and Circles

In This Chapter

- Defining conic sections
- General and standard forms of a conic
- Identifying the vertex, focus, directrix, and axis of symmetry of a parabola
- Identifying the center and radius of a circle
- Graphing parabolas and circles in the coordinate plane

Quadratic equations are old news. You already know how to solve them (using the quadratic formula, completing the square, or factoring), and you have even drawn a few of their graphs (by plotting points or using graph transformations). However, we've barely begun to scratch the surface of quadratic graphs (called *parabolas*). Have no fear—once this chapter is over, the surface will be scratched up pretty severely. Surprisingly, there's a ton of stuff to learn about parabolas, although it may feel weird working so hard to understand graphs you were already drawing way back in Chapter 6.

Once we get cooking with quadratic equations, it'll be time to up the ante and toss two quadratic expressions into the *same* equation. It sounds kind of complicated, but its graph (a circle) couldn't be simpler.

These two seemingly unrelated shapes aren't grouped in the same chapter coinciden-tally; they're two of a total of four graphs that together make up something called the *conic sections*. (I discuss the other two graphs in Chapter 16.)

Introducing Conic Sections

The *conic sections* are a group of four geometric shapes: parabolas (the *u*-shape graphs of quadratic equations), circles, ellipses (ovals), and hyperbolas (which you may never have heard of in your entire life, but they sort of look like two parabolas stuck to-gether). They are called the conic sections because you can generate their shapes by slicing into a right circular cone. Oddly, this is also how jazz musician Harry Connick Jr. got his name.

Talk the Talk

The **conic sections** con-sist of four geometric shapes (parabolas, circles, ellipses, and hyperbolas) that are actually the cross-sections of a right circular cone sliced by a plane.

Not many people have experience with cones (unless you build roads or work at Dairy Queen), so check out Figure 15.1 to visualize what I'm talking about. A right circular cone is nothing more than a cylinder that gets gradually thinner and eventually tapers off to a point. It's called a "right" cone because it stands straight up and down rather than slanting to one side.

Figure 15.1

Slicing parallel to the cone's base gives you a circle, but if you come in at an angle, you end up with an ellipse.

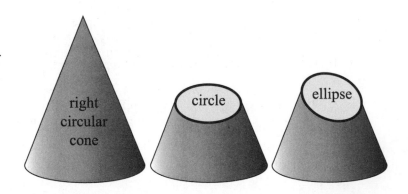

right circular cone

circle

ellipse

Imagine using a sharp knife to slice that cone at the far left of Figure 15.1. If you cut it straight across its middle, moving the blade exactly parallel to the base of the cone, you end up with a perfectly circular cross-section. However, cutting at an angle results in a cross-sectional that's elliptical. (To get parabolas and hyperbolas, you slice off a side of the cone rather than cutting through the middle.)

As fun as it is to imagine slicing mathematical things with a giant knife, that maniacal look in your eye is frightening me, so I'll return to the less dangerous, less sharp, less "Hey, I wonder what *your* cross-section would look like" side of things, and back into the safely dull land of equations.

Every conic section can be written in *general form*, which looks like this: $Ax^2 + By^2 + Cx + Dy + E = 0$. Notice that conics may contain as many as two different squared variables (which is new), and the general form requires you to set everything equal to 0 (which isn't new).

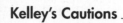

Kelley's Cautions

Very advanced conic sections written in general form will also contain an xy term, which indicates that its graph is rotated in the coordinate plane. However, you shouldn't see any of those in your precalculus class, so I don't discuss them here.

Even though you don't know a lot about how the individual conic sections work just yet, you can still tell them apart. To classify a conic section in general form, $Ax^2 + By^2 + Cx + Dy + E = 0$, just look at the coefficient(s) of the squared term or terms (A and B). If either A or $B = 0$ (in other words, there's only one squared term in the equation), you've got a parabola on your hands. However, if both A and B are nonzero numbers, one of three possibilities exists:

◆ If $A = B$, it's a circle.

◆ If $A \neq B$, but A and B both have the same sign, it's an ellipse.

◆ If $A \neq B$, and A and B have different signs, you've got a hyperbola.

Once you can classify conics, you're ready to try the two major tasks expected of you for every conic section: rewriting equations in standard form and graphing them.

Example 1: Identify the conic section represented by each of the following equations in general form:

a. $3x^2 + 3y^2 - 5x - 9y + 1 = 0$

Solution: The coefficients of x^2 and y^2 are equal—they're both 3—so this is the equation of a circle.

b. $2x^2 + y^2 + 6x - 2y - 8 = 0$

Solution: This time, the coefficient of x^2 (2) doesn't equal the coefficient of y^2 (1), so it's not a circle. However, both coefficients are positive, so it must be an ellipse.

c. $5x^2 - 5y^2 + 6x - y + 3 = 0$

Solution: The coefficients of x^2 and y^2 (5 and -5, respectively) are unequal and have different signs. Therefore, this must be the equation of a hyperbola.

You've Got Problems

Problem 1: Identify the conic section represented by each of the following equations in general form:

   a. $3x^2 - y^2 + 7x - 7y - 2 = 0$
   b. $5y^2 + x - 2y = 0$

Polishing Off Parabolas

Back in Chapter 6 (in the section "Function Transformations"), you learned how to graph a function like $f(x) = 2(x + 1)^2 - 3$; it's just the graph of $y = x^2$ moved left one unit and down three units, and stretched a bit vertically. (Those were innocent times, weren't they? Such good memories.) What you didn't know at the time was that you were actually dealing with the standard form of a quadratic equation, whose graph is a parabola.

Before we get started, I should give you the fancy-pants mathematical definition of a parabola. A *parabola* is the set of points that are equidistant (the same distance) from a point (called the *focus*) and a line (called the *directrix*).

Talk the Talk

A **parabola** is the set of points on the coordinate plane that are equidistant from a fixed point (called the **focus**) and a fixed line (called the **directrix**).

Imagine this scene: A straight road runs near your house, and about 30 feet away from that road, you have a tall tree in your yard. If you and all of your friends who enjoy trees, roads, parabolas, and strange math illustrations gathered together on the same side of the road as the tree, and each person stood the same distance away from the tree and the road, the group of you would form a parabolic graph. (If you're having a hard time picturing this, check out Figure 15.2; the tree is the focus and the road is the directrix.)

You'll see two different kinds of parabolas in precalculus, depending upon which of the squared terms (either x^2 or y^2) is missing from its general form. Remember, the equation of a parabola will have either an x^2 term or a y^2 term, but not both. If a parabola contains an x^2 term, its graph points either up or down, and a parabola containing a y^2 term points either left or right.

Quadratics Containing an x^2 Term

If a parabola contains an x^2 term, it looks like this in standard form:

$$y = a(x - h)^2 + k, \text{ where } a = \pm\frac{1}{4c}$$

Hold on a minute—I think I heard your heartbeat suddenly triple. I know that equation looks really bizarre, but it makes more sense if you look at it in terms of its graph, shown in Figure 15.2.

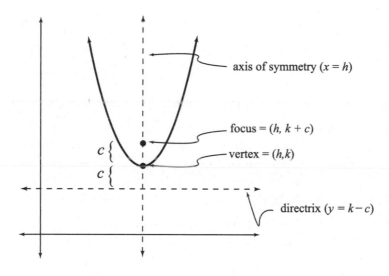

Figure 15.2

All of the values marked on this parabola are accurate if the parabola points upward. If it points down, the focus will be (h, k – c) *and the directrix will be* y = k + c.

Let's go through the major parts of the parabola so you can see what all of the variables a, c, h, and k mean and where they come from:

♦ The *vertex* of the parabola—its lowest point if the graph points up and its highest point if the graph points down—is located at point (h,k).

♦ The *axis of symmetry* is a vertical line that passes through the vertex, so it has the equation $x = h$. It cuts the graph into two equal halves that look like reflections of one another across the line; technically, this means the parabola is symmetric about that axis, which is where the name comes from.

◆ Every point on the parabola is the same distance from the focus and directrix, but since the vertex lies on the axis of symmetry, it's easiest to calculate. The distance from the vertex to each of those landmarks is labeled c.

Talk the Talk

The **axis of symmetry** is a line that cuts through the middle of the parabola, intersecting at only one point, called the **vertex**. The distance from the vertex to both the focus and directrix (measured along the axis of symmetry) is the value c in the standard form equation $\left(a = \pm\dfrac{1}{4c}\right)$.

◆ When the parabola points up, its focus is above its vertex, at the point $(h, k + c)$. Notice that the x-value, h, is the same. Both are located on the axis of symmetry, $x = h$. If the parabola points down, the focus has the coordinates $(h, k - c)$.

◆ The directrix is always located on the side of the vertex opposite the focus, a horizontal line exactly c units below an upward-pointing parabola ($y = k - c$) or c units above a parabola that points down ($y = k + c$).

To convert a parabola from general to standard form, you need to complete the square. If you can't remember how that works, flip back to Chapter 7 and review it before you try the following example. (That troublesome binomial square, $(x - h)^2$, makes completing the square necessary.)

If you absolutely *hate* completing the square, I have some bad news for you. Every conic section contains at least one binomial square. In fact, the other three conics you've yet to deal with will have *two* binomial squares each, so you'll have to complete the square *twice* in each of those problems to reach standard form. If this really freaks you out, just remember that things could always be worse. What if your shirt caught on fire while you were trying to put a parabola in standard form? That would be *much* worse.

Critical Point

If the focus of a parabola is higher than the directrix in the coordinate plane, the parabola must point up (toward the focus and away from the directrix). On the other hand, if the directrix is above the focus, the parabola must point down.

Example 2: Write the equations of the parabolas in standard form and graph them:

a. Parabola with focus $(-2,4)$ and directrix $y = 5$

Solution: The axis of symmetry passes through both the focus and the vertex, so their x-values must match. Therefore, the equation of the axis of symmetry must be $x = -2$, and the x-coordinate of the vertex (h) is also -2. Also notice that the focus is below the directrix, so the parabola will point down.

To calculate c, find the length of the vertical segment connecting the focus and directrix (subtract their heights from one another and take the absolute value), and divide that number by 2. In this problem, the height of the focus is 4 and the height of the directrix is 5. Therefore, they're $|4-5|=1$ units apart, and $c = \frac{1}{2}$. Count c units from the focus toward the directrix—in this case, you count $\frac{1}{2}$ unit up from (–2,4) to reach the vertex: $\left(-2, 4 + \frac{1}{2}\right) = \left(-2, \frac{9}{2}\right) = (h, k)$.

At this point, you know $h = -2$ and $k = \frac{9}{2}$, but you don't know what a is. Fortunately, you do know that $a = \pm\frac{1}{4c}$, and since you just figured out that $c = \frac{1}{2}$, plug it into the formula:

$$a = \pm\frac{1}{4\left(\frac{1}{2}\right)} = \pm\frac{1}{2}$$

Be careful! You haven't actually finished calculating a; you know its value, but you don't know its sign. If the graph faces up, a must be positive, but if the graph faces down, a must be negative. As I mentioned earlier, this parabola points down, so $a = -\frac{1}{2}$.

Plug a, h, and k into standard form, and you've got the equation of the parabola:

$$y = a(x - h)^2 + k$$
$$y = -\frac{1}{2}(x - (-2))^2 + \frac{9}{2}$$
$$y = -\frac{1}{2}(x + 2)^2 + \frac{9}{2}$$

CAUTION **Kelley's Cautions**

The constant c will always be positive, so the formula $a = \frac{1}{4c}$ cannot tell you whether a should be positive or negative. The sign of a depends on the direction the parabola faces.

Figure 15.3 shows the graph of this parabola. It's just the graph of $y = x^2$ with a few transformations applied to it: Reflect it across the x-axis, flatten it out just a little so that each height is $a = \frac{1}{2}$ as high as it started out, and finally move it left 2 units and up $\frac{9}{2}$ units.

Figure 15.3

Although you can use transformations to graph $y = -\frac{1}{2}(x + 2)^2 + \frac{9}{2}$, *it's just as easy to draw the graph by plotting points.*

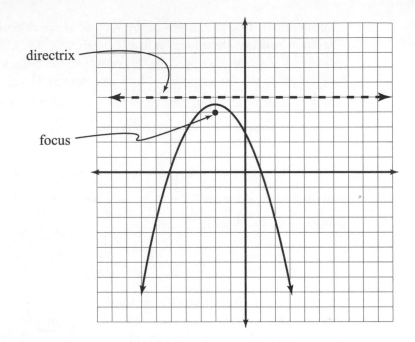

directrix

focus

Critical Point

A parabolic equation solved for *y* is not completely useless. The constant is also the *y*-intercept of the graph.

Kelley's Cautions

Be careful when completing the square for a quantity inside parentheses—like in Example 2b. If you add something inside those parentheses, you have to add that number *multiplied by the outer constant* to the other side of the equation.

b. Parabola with equation $y = 2x^2 + 12x + 22$

To force this equation into standard form, you need to complete the square. Start by moving the constant to the opposite side of the equation:

$$y - 22 = 2x^2 + 12x$$

Don't forget that the x^2 coefficient must be equal to 1, or completing the square won't work. Instead of dividing everything by 2, factor that coefficient out of both terms on the left side. (This is where the *a* term in standard form comes from.)

$$y - 22 = 2(x^2 + 6x)$$

Take half of *x*'s coefficient (half of 6 equals 3), and square it (3^2) to get 9. This means you should add 9 inside the parentheses, right after the 6x. To keep the equation correctly balanced, you need to add the same quantity to the left side as well, but be careful!

Even though it looks like you added 9 to the right side of the equation, you actually added 9 inside a group of parentheses multiplied by 2. In other words, you actually added $9 \cdot 2 = 18$ to the right side, so you must add it to the right side as well:

$$y - 22 + 18 = 2(x^2 + 6x + 9)$$

Simplify the left side of the equation and factor the right side:

$$y - 4 = 2(x + 3)^2$$

Solve for y (by adding 4 to both sides), and the parabola's in standard form:

$$y = 2(x + 3)^2 + 4$$

Compare this to the chock-full-of-variables standard form of a parabola, $y = a(x - h)^2 + k$, to see that $a = 2$ (which is positive, so the parabola faces up), $h = -3$ (h is always the *opposite* of the constant in parentheses), and $k = 4$. Generate the graph in Figure 15.4 by either plotting points or applying transformations.

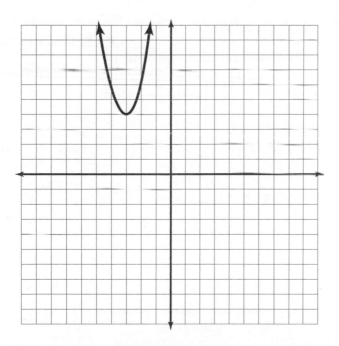

Figure 15.4

The upward-pointing graph of y = 2(x + 3)² + 4 *has the vertex* (−3, 4)

You've Got Problems

Problem 2: Identify the focus and vertex of the parabola with the equation $y = 4x^2 - 8x + 7$.

Quadratics Containing a y^2 Term

Whereas circles, ellipses, and hyperbolas must always contain both an x^2 term and a y^2 term, parabolas can contain only one or the other. Until now, every parabola I've thrown at you has been of the x^2 variety, so it's time to examine some y^2 ones. Although they have a lot in common, there are two key differences:

1. **Instead of pointing upward or downward (like x^2 parabolas), y^2 parabolas point right or left.** You can still tell which direction it's pointing by looking at the sign of a. If a is positive, the parabola's pointing right; a negative a indicates a parabola pointing left.

2. **They are not functions.** All y^2 parabolas (like the one pictured in Figure 15.5) fail the vertical line test because just about every point on the parabola has a sister point with the same x-value. In other words, lots of x-values have *two* matching y-values; for those x inputs, there are two outputs, and this breaks the rules a function must abide by.

Although those are significant differences, their standard forms are remarkably similar. The standard form of a parabola containing a y^2 term looks like this: $x = a(y - k)^2 + h$, if $a = \pm\dfrac{1}{4c}$; its graph will look something like Figure 15.5.

Figure 15.5

The graph of a parabola containing a y^2 looks a lot like the graph of an x^2 parabola rotated 90° clockwise.

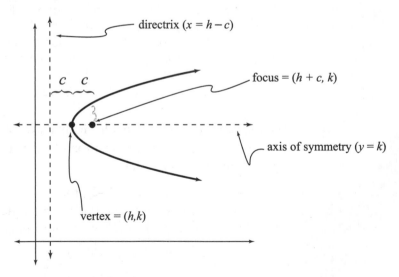

Spend a moment acquainting yourself with the graph in Figure 15.5 and investigating the important characteristics of the now-horizontal parabola:

- The equation is solved for *x* rather than *y*, and the squared binomial now houses the *y* term.

- The vertex is still (*h,k*), although the *h* and *k* appear in different places in standard form.

- While *c* still represents the distance from the vertex to both the focus and the axis of symmetry, it is now a horizontal distance rather than a vertical one.

- Speaking of horizontal, the axis of symmetry also heads east and west rather than north and south; therefore, its equation now begins with *y* = instead of *x* =.

- The orientation of the directrix has changed; it's now a vertical line, with an equation that begins with *x* =.

The most important thing to remember is that *c* is now a horizontal distance, meaning you'll move left and right from the vertex if you're trying to reach the focus and directrix. Therefore, you should add or subtract *c* from *h* (instead of *k*).

Critical Point _____

Have you ever heard of a "parabolic microphone"? Among other things, they're often used to capture on-field sounds along the sidelines at a football game. Satellite dishes and flashlight reflectors have a parabolic cross-section as well, for the same reason: Sounds, light rays, and satellite beams hitting a parabolic surface all reflect back to one point, the focus of the parabola, which concentrates and intensifies the incoming signal and thereby improves its quality.

Example 3: Assume that a parabola with a horizontal axis of symmetry has vertex (4,–2) and *y*-intercept at (0,3). Write the equation of the parabola in standard form, identify its focus and directrix, and graph it.

Solution: If the vertex of the parabola is (4,–2), then *h* = 4 and *k* = –2. That's fantastic information because when they're substituted into the standard form for a parabola with a horizontal axis of symmetry, the only constant remaining unknown is *a*:

$$x = a(y-k)^2 + h$$
$$x = a(y+2)^2 + 4$$

Now plug in the coordinates of the y-intercept ($x = 0$, $y = 3$) and solve for a:

$$0 = a(3+2)^2 + 4$$
$$-4 = 25a$$
$$a = -\frac{4}{25}$$

So, the equation of the parabola in standard form is $x = -\frac{4}{25}(y+2)^2 + 4$. Use the formula $a = \frac{1}{4c}$ to calculate c (ignore the sign of a; c is always positive):

$$\frac{4}{25} = \frac{1}{4c}$$
$$\frac{4}{25} = \frac{1}{4c}$$
$$16c = 25$$
$$c = \frac{25}{16}$$

So, you must travel $\frac{25}{16}$ units from the vertex to reach either the focus or the directrix. Remember, the sign of a tells you which way the parabola faces and, therefore, which direction you go from the vertex to reach the focus. In this problem, a is negative, so the parabola must face left. Subtract $\frac{25}{16}$ from the x-coordinate of the vertex to calculate the focus:

$$\left(4 - \frac{25}{16}, -2\right) = \left(\frac{64}{16} - \frac{25}{16}, -2\right)$$
$$\text{focus} = \left(\frac{39}{16}, -2\right)$$

Critical Point

You could always graph the parabola by plotting a bunch of points as well. Just plug a bunch of y-values into $x = -\frac{4}{25}(y+2)^2 + 4$ (preferably, y-values that are close to the vertex's y-value of −2), and calculate the corresponding x-values to get points on the graph.

To determine the equation of the directrix, add $\frac{25}{16}$ to the x-value of the vertex. (Remember that the directrix of a parabola that points in a horizontal direction must be a vertical line and will have the equation $x =$.)

$$x = 4 + \frac{25}{16} = \frac{64}{16} + \frac{25}{16} = \frac{89}{16}$$
$$\text{directrix: } x = \frac{89}{16}$$

The easiest way to graph the parabola is to plot the vertex and y-intercept as I've done in Figure 15.6. Since each point on the parabola must have a corresponding sister point reflected across the axis of symmetry, the point $(0,-7)$ also belongs on the graph. (It, like the y-intercept $(0,3)$, lies on the y-axis and is exactly 5 vertical units away from the axis of symmetry.) Use those three points to get a rough idea of the graph's shape.

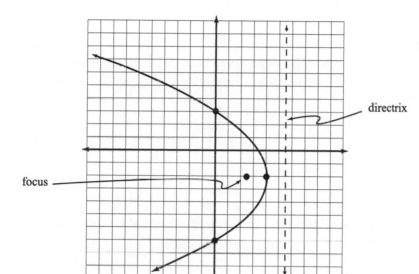

Figure 15.6

You don't have to draw the focus and directrix when graphing a parabola, as I have done here for the graph of $x = -\frac{4}{25}(y+2)^2 + 4$, *but I just couldn't help myself. Once I get to graphing, nothing can stop me.*

directrix

focus

You've Got Problems

Problem 3: Put the equation of the parabola $x = y^2 - 6y + 8$ in standard form; calculate its vertex, focus, and directrix; and draw its graph.

The Vicious Circle

Believe it or not, that little crash course you just got in parabolas is great preparation for the rest of the conic sections. There are only two tricky skills to master when dealing with conics: memorizing the standard forms of each one and successfully completing the square without weeping. Once you can do both, conic sections are a breeze. If you don't quite get the hang of them yet, don't worry. Not only will they get easier with practice, but I think you'll find that the pages in this book are surprisingly absorbent, so your tears will soak right in and smudge the ink only a little.

Finally, some good news. *Circles* are much, much, much, much easier to understand than parabolas, for a number of reasons:

- There are only two major features of a circle, the *center* and *radius*, as opposed to a parabola's more numerous features (vertex, focus, directrix, axis of symmetry).

- A circle has only one standard form, unlike the two standard forms of a parabola.

- Circles are incredibly simple to graph—even easier than lines! In fact, the only thing easier to graph than a circle is a single point.

I know you're itching to find out, so here's the equation for the standard form of a circle:

$$(x - h)^2 + (y - k)^2 = r^2$$

When you graph that, you end up with something that looks like Figure 15.7. Notice that (h,k) again represents an important point, like it did in parabolas, but instead of a vertex, it now represents the center of the circle. You probably already guessed it, but I'll tell you anyway. The variable r represents the circle's radius.

Figure 15.7

The graph of a circle in standard form, $(x - h)^2 + (y - k)^2 = r^2$, where (h,k) is the center and r is the radius.

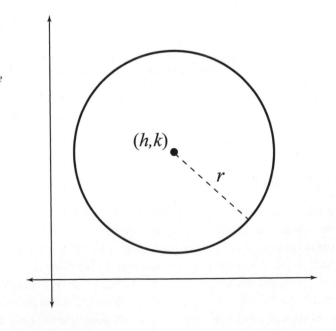

(h,k)

r

Example 4: Put the equation of the circle $4x^2 + 4y^2 - 8x + 12y - 23 = 0$ in standard form, identify its center and radius, and graph it.

Solution: You can be sure that this is a circle because its x^2 and y^2 coefficients match. That said, go ahead and divide *everything* by 4 to cancel out those coefficients so you can complete the square for both x and y:

$$x^2 + y^2 - 2x + 3y - \frac{23}{4} = 0$$

Since you're going to complete the square twice, it helps to visualize the left side as two different groups, one containing x variables and one containing y variables. Move the constant across the equals sign as well, by adding it to both sides:

$$\left(x^2 - 2x\right) + \left(y^2 + 3y\right) = \frac{23}{4}$$

Now complete the square for the x terms (by adding 1 in the leftmost set of parentheses) and for the y terms (by adding $\frac{9}{4}$ inside the other parentheses). Don't forget to keep the equation balanced by adding these values to the right side of the equation as well:

$$\left(x^2 - 2x + 1\right) + \left(y^2 + 3y + \frac{9}{4}\right) = \frac{23}{4} + 1 + \frac{9}{4}$$

Factor the trinomials and add up that ugly string of constants on the right side; you end up with standard form:

$$\left(x - 1\right)^2 + \left(y + \frac{3}{2}\right)^2 = 9$$

In this equation, $h = 1$ and $k = -\frac{3}{2}$ (the opposites of the numbers inside parentheses), so the center of the circle is $\left(1, -\frac{3}{2}\right)$. You also know that $r^2 = 9$, so the radius of the circle is $3\left(\sqrt{r^2} = \sqrt{9} = 3\right)$.

To graph the circle, plot the center and then count 3 units (the length of the radius) up, down, right, and left of the center, marking each point. Finally, draw a circle passing through those four points. You end up with Figure 15.8.

Kelley's Cautions

Don't forget that the standard form of a circle contains the radius *squared*. You have to take the square root of the lone constant from the right side of the equation to figure out what the radius is.

Figure 15.8

The graph of

$$(x-1)^2 + \left(y + \frac{3}{2}\right)^2 = 9, \text{ the}$$

circle that looked like $4x^2 + 4y^2 - 8x + 12y - 23 = 0$ in general form.

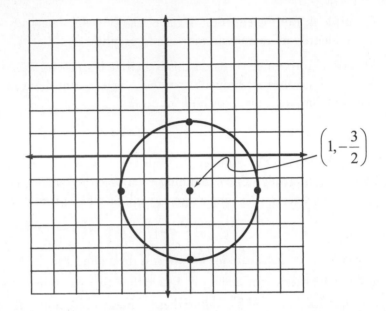

$$\left(1, -\frac{3}{2}\right)$$

You've Got Problems

Problem 4: Put the equation of the circle $x^2 + y^2 + 6x - 10y + 30 = 0$ in standard form, identify its center and radius, and graph it.

The Least You Need to Know

- There are four kinds of conic sections: parabolas, circles, ellipses, and hyperbolas.

- You can classify a conic section written in general form using the coefficients of its x^2 and y^2 terms.

- A parabola is the set of points in the coordinate plane that are equidistant from a fixed point and a fixed line. Its standard form is either $y = a(x - h)^2 + k$ or $x = a(y - k)^2 + h$, where $a = \pm\dfrac{1}{4c}$.

- A circle is the set of points in the coordinate plane that are equidistant from a fixed point. Its standard form is $(x - h)^2 + (y - k)^2 = r^2$.

Chapter 16

Ellipses and Hyperbolas

In This Chapter

♦ Converting equations of ellipses and hyperbolas from general to standard form

♦ Graphing ellipses and hyperbolas that aren't necessarily centered at the origin

♦ Calculating and interpreting the eccentricity of an ellipse

Consider a kernel of unpopped popcorn. (That sounds very Zen, doesn't it?) A cross-section of the kernel wouldn't be circular, but rather would have an oval (or elliptical) shape. The forces that shape the kernel (I'm no scientist, so they might be some or all of the following: gravity, inertia, centrifugal force, centripetal force, surface tension, hydrostatic force, fission, fusion, photosynthesis, cytoplasmic force, ectoplasmic force, or plasmaplasmic force) focus the contents inward. However, when heated, the moisture in the kernel expands, eventually blowing the kernel apart in a massive outward focus of force, a blast so powerful it can be contained only by a greasy, salty, microwavable bag. The unpopped and popped kernels look dramatically different (as I assume we all would if our moisture exploded out of our skin at terrific speeds), but all of the parts are still there. They're just sort of inside out once the kernel pops.

If an ellipse is similar to an unpopped kernel, a hyperbola is what you get when the ellipse is heated up and blown apart. As you'll find out in this chapter, they basically have all of the same pieces; the only difference is where the energy is focused: inward (in an elliptical kernel) or outward (in a popped hyperbola).

Eclipsing Ellipses

The third conic section is the *ellipse*, which (as I've mentioned) looks like an oval and (like a popcorn kernel) hurts your teeth if you chew on it. It gets its shape in a way vaguely similar to a circle. However, in a circle, each point is exactly the same distance away from one fixed point; an ellipse, on the other hand, is formed by two fixed points, each one called a *focus* (or *foci*, if you're talking about both of them at once).

If you take any point on an ellipse and add the two distances measured from that point to each of the foci, you'll end up with the same total. Check out Figure 16.1 to see what I mean.

In Figure 16.1, point X on the ellipse is a distance of a away from focus F_1 and a distance of b away from the other focus, F_2. The sum of those distances ($a + b$) will match the sum of the focal distances for any other point on the ellipse. Therefore, $c + d = a + b$, even though points X and Y are at completely different locations on the ellipse.

Talk the Talk

An **ellipse** is the set of all points on the coordinate plane such that the sum of the distances from each point to two fixed points (called the **foci**) remains constant.

Critical Point

Here's an interesting bit of trivia for you: The orbits of the planets around the sun are elliptical, not circular, like most plastic models of the solar system (made of rotating wires and plastic balls) may lead you to believe.

Figure 16.1

This ellipse is formed by the two foci, labeled F_1 *and* F_2; *points* X *and* Y *lie on the ellipse itself. The constants* a, b, c, *and* d *represent the distances between* X *and* Y *and the foci.*

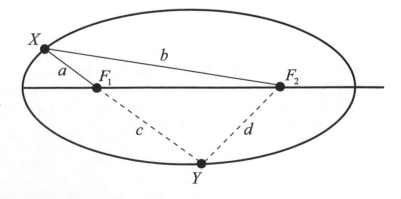

Elliptical Anatomy

There's more to an ellipse than just its foci (it also has a wonderful personality and fascinating talents like juggling), so let me break down a generic ellipse into its component parts for you. In Figure 16.2, I've drawn two ellipses, one that stretches horizontally and one that stretches vertically. These are the only two kinds of ellipses we'll discuss. Others, rotated at jaunty angles, are too advanced for precalculus.

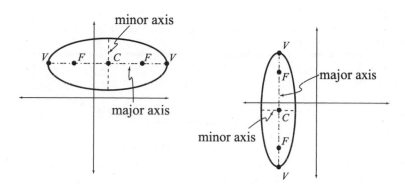

Figure 16.2

In these ellipses, the foci are labeled F, *the vertices are labeled* V, *and the center points are labeled* C. *Notice that the major axis is always longer than the minor axis.*

If you connect the foci of an ellipse, the midpoint of the segment you create is called the *center* of the ellipse. Through the center pass two perpendicular axes, one horizontal and one vertical. The longer of the two axes is called the *major axis*, and the shorter is the *minor axis*. Each endpoint of the major axis is called a *vertex* of the ellipse (use the plural form, vertices, when you're talking about them both).

In an ellipse with a horizontal major axis, the points representing the foci, the vertices, and the center all have the same *y*-value because they're all located on the same horizontal line segment. Similarly, all five of those points have the same *x*-value on an ellipse with a vertical major axis.

Talk the Talk

The midpoint of the segment whose endpoints are the foci of an ellipse is called the ellipse's **center.** Two perpendicular axes pass through the center and have endpoints on the ellipse. The longer of the two is called the **major axis,** and the shorter is the **minor axis.** Each endpoint of the major axis is called a **vertex** of the ellipse.

Standard Form of an Ellipse

All those axes, foci, center points, and vertices translate into variables that appear in the standard form for an ellipse, as illustrated in Figure 16.3.

Figure 16.3

An ellipse centered at the point (h,k) with major axis length 2a and minor axis length 2b. Even if the major axis of this ellipse were vertical rather than horizontal, the major axis would be written in terms of a *and the minor axis in terms of* b.

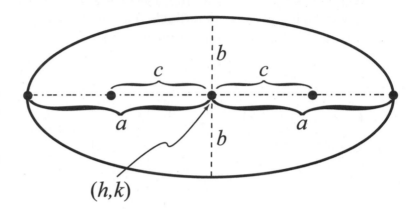

Here's what all the constants in Figure 16.3 mean:

♦ **The center of the ellipse is the point (*h*,*k*).** This is the same ordered pair you used for the center of a circle.

♦ **The length of the major axis is 2*a*.** In other words, you'll travel exactly *a* units right and left from the center to reach the ellipse's vertices (if the major axis is horizontal, like in Figure 16.3; otherwise, you'll travel *a* units up and down from the center).

Critical Point _____

Did you know that a circle is technically just a special kind of ellipse, just like a square is a special kind of rectangle? A circle is just an ellipse whose center and foci are all the exact same point.

♦ **The length of the minor axis is 2*b*.** Just as with the major axis, you must travel *b* units in opposite directions to reach the ellipse along the minor axis.

♦ **The distance from the center to each focus is *c*.** This is similar to a parabola, in which the distance *c* represents the distance from the vertex to both the focus and the directrix.

There are two standard forms of an ellipse:

$$\frac{(x-h)^2}{a^2}+\frac{(y-k)^2}{b^2}=1 \quad \text{or} \quad \frac{(x-h)^2}{b^2}+\frac{(y-k)^2}{a^2}=1 \text{, if } c=\sqrt{a^2-b^2}$$

The only real difference between them is their denominators. Once you're finished putting an elliptical equation in standard form, you label the larger denominator a^2 and the smaller one b^2; that's all there is to it. Here's the impact of that simple difference: If the larger denominator is located beneath $(x - h)^2$, the ellipse has a horizontal major axis, but if $(y - k)^2$ has the larger denominator, the ellipse's major axis is vertical. That's easy to remember: If the x denominator is bigger, think of the x-axis (which is horizontal); if the y denominator is bigger, think of the y-axis (which is vertical).

Since there are two binomial squares in the formula (just as in the standard form of a circle), you have to complete the square twice to reach standard form, at least in most cases. Unlike circles, however, an ellipse's standard form is not set equal to r^2. In fact, it's not set equal to a variable at all. An ellipse in standard form *always* equals 1.

Example 1: Write the equation of each ellipse in standard form and graph it:

a. The ellipse with vertices (6,1) and (6,–9), and foci (6,0) and (6,–8)

   Solution: Since the vertices and the foci all fall on the same vertical line ($x = 6$), the major axis is vertical; the center of the ellipse is the midpoint of the segment connecting the foci: (6,–4).

   To put the ellipse in standard form, you need values for h, k, a, and b. At this point, you already know that $h = 6$ and $k = -4$ (the coordinates of the center), and you can easily calculate a (the distance from the center to either vertex) and c (the distance from the center to either focus): $a = 5$ and $c = 4$. Plug the values of a and c into the formula $c = \sqrt{a^2 - b^2}$ to calculate b:

   $$4 = \sqrt{5^2 - b^2}$$
   $$(4)^2 = \left(\sqrt{5^2 - b^2}\right)^2$$
   $$16 = 25 - b^2$$
   $$-9 = -b^2$$
   $$b = \pm 3$$

> **Critical Point**
>
> If the midpoint of a segment isn't obvious, use the midpoint formula from geometry: A segment with endpoints (x_1, y_1) and (x_2, y_2) has midpoint $\left(\dfrac{x_1 + x_2}{2}, \dfrac{y_1 + y_2}{2}\right)$. In other words, the x-value of the midpoint is the average of the endpoints' x-values, and the y-value works the same way.

The sign of b (like the sign of a) doesn't really matter because it's squared in standard form anyway, although most people like writing a and b as positive numbers, since they represent distances. Now plug $h = 6$, $k = -4$, $a = 5$, and $b = 3$

into the standard form of an ellipse. In this case, a^2 should be written beneath $(y-k)^2$, since the major axis is vertical:

$$\frac{(x-b)^2}{b^2}+\frac{(y-k)^2}{a^2}=1$$

$$\frac{(x-6)^2}{9}+\frac{(y+4)^2}{25}=1$$

Kelley's Cautions

Don't connect the four points on the graph with straight lines. You're graphing an ellipse, not a diamond!

To graph this equation, plot the center point and the vertices given to you in the original problem. Since $b = 3$ (and the minor axis is horizontal), count 3 units left and right of the center, and plot those points as well. Connect the four endpoints of the major and minor axes using an oval, like in Figure 16.4.

Figure 16.4

The graph of the ellipse $\dfrac{(x-6)^2}{9}+\dfrac{(y+4)^2}{25}=1$, *with major axis length* 2a = 10 *and minor axis length* 2b = 6.

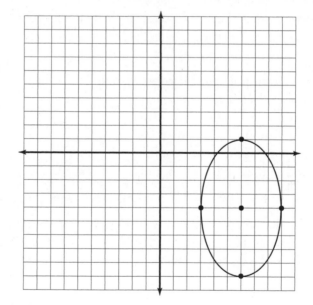

b. $16x^2 + 36y^2 + 32x - 216y - 236 = 0$

Solution: Begin by grouping the terms so that the x terms and the y terms are next to one another in descending power order; move the constant to the right side of the equation:

$$(16x^2 + 32x) + (36y^2 - 216y) = 236$$

Factor the x^2 coefficient out of the two x terms only, and then factor the y^2 coefficient out of the two y terms:

$$16(x^2 + 2x) + 36(y^2 - 6y) = 236$$

Complete the square twice by adding 1 inside the x parentheses and 9 inside the y parentheses. To keep the equation balanced, you should also add those numbers to the right side *after you multiply them by the corresponding numbers in front of the parentheses.* In other words, you add $16(1) = 16$ and $36(9) = 324$ to the right side of the equation:

$$16(x^2 + 2x + 1) + 36(y^2 - 6y + 9) = 236 + 16 + 324$$

Factor and simplify:

$$16(x + 1)^2 + 36(y - 3)^2 = 576$$

An ellipse in standard form *must* be equal to 1, so divide all three terms by 576 to cancel out that gigantic constant. Then simplify the fractions:

$$\frac{16(x+1)^2}{576} + \frac{36(y-3)^2}{576} = \frac{576}{576}$$

$$\frac{(x+1)^2}{36} + \frac{(y-3)^2}{16} = 1$$

Since $36 > 16$, the major axis must be horizontal (36 is the denominator of the fraction containing x). Therefore, $a = \sqrt{36} = 6$, $b = \sqrt{16} = 4$, $h = -1$, and $k = 3$. (Don't forget that h and k are always the opposite of the constants in the squared binomials.) This means that the ellipse is centered at $(-1,3)$, has a horizontal major axis of length $2a = 12$, and has a vertical minor axis of length $2b = 8$.

To generate the graph (pictured in Figure 16.5) count 6 units left and right of the center, and plot the endpoints of the major axis. Then count 4 units up and down from the center to plot the minor axis endpoints. Connect all four points using an oval.

Figure 16.5

The almost-track-and-field-shape graph of $\dfrac{(x+1)^2}{36} + \dfrac{(y-3)^2}{16} = 1$. *You can almost imagine that Marion Jones is one of the vertices.*

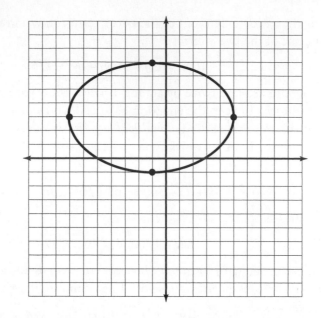

You've Got Problems

Problem 1: Write the elliptical equation $x^2 + 4y^2 - 4x + 40y + 100 = 0$ in standard form, graph it, and calculate the length of its major axis.

Calculating Eccentricity

All people, deep down, possess some level of eccentricity. I'm not necessarily talking about the debilitating craziness some people demonstrate via their collection of 650 cats in a two-bedroom apartment or a wardrobe consisting only of replica Civil War uniforms and a dogged insistence that everyone address you as "Colonel." Even so, everyone has their little foibles. Some are more pronounced than most.

Talk the Talk

The **eccentricity** of an ellipse, defined as $\dfrac{c}{a}$, is a value on the interval $[0,1)$ that describes the "ovalness" of an ellipse.

When you discuss the *eccentricity* of an ellipse, you're not talking about how crazy it is; you're describing its "ovalness." Just like human craziness, some ellipses will have a more pronounced eccentricity than others. In the case of ellipses, you're answering the question "Does this ellipse look more like an oval or more like a circle?" using this simple formula: eccentricity $= \dfrac{c}{a}$.

In other words, the eccentricity of an ellipse is equal to c (the distance from the center to a focus) divided by a (the distance from the center to a vertex). The closer to 0 the eccentricity is, the more the ellipse resembles a circle; in fact, an ellipse with an eccentricity of exactly 0 actually *is* a circle. The larger the eccentricity is, the more pronounced oval shape the ellipse has.

Example 2: Calculate the eccentricity of the ellipse with the equation

$$\frac{(x-6)^2}{14} + \frac{(y-9)^2}{64} = 1$$

Solution: This equation is already in standard form, so $h = 6$, $k = 9$, $a^2 = 64$, and $b^2 = 14$. To calculate eccentricity, you need to know what c equals, so use the formula $c = \sqrt{a^2 - b^2}$:

$$c = \sqrt{64 - 14}$$
$$c = \sqrt{50} = 5\sqrt{2}$$

The eccentricity is simply c divided by a:

$$e = \frac{5\sqrt{2}}{\sqrt{64}} \approx 0.884$$

Since 0.884 is closer to 1 than 0, the ellipse is more oval than circle.

You've Got Problems

Problem 2: Calculate the eccentricity of the ellipse with the equation $5x^2 + y^2 - 3y + 1 = 0$.

Handling Hyperbolas

A *hyperbola*, as I mentioned in my delicious popcorn metaphor at the beginning of the chapter, looks like an ellipse blown inside out. Whereas the sum of the distances from a point on an ellipse to the two foci was guaranteed constant, a hyperbola does the exact opposite. It guarantees that the *difference* of those distances will always be constant. As you can see in Figure 16.6, that one little change in the definition has a gigantic impact on the poor, unsuspecting elliptical graph.

Figure 16.6

In these graphs of hyperbolas with horizontal and vertical transverse axes, the variables stand for the same things they did with ellipses: V = *vertex,* F = *focus, and* C = *center.*

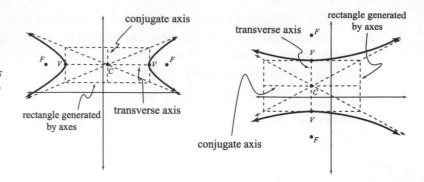

Like an ellipse, a hyperbola has foci, vertices, and a center; it also has two perpendicular axes that chop through one another right at the center point, but they've got different names. The actual lengths of those axes are not quite as important as they were in an ellipse, so you won't refer to them using the length-based descriptors "major" and "minor." Instead, the axis that connects the vertices is now called the *transverse axis*, and the other is called the *conjugate axis*.

Talk the Talk

If you measure the distances from any point on a **hyperbola** to two fixed points (called the **foci**) and calculate the difference, you get the same constant. The midpoint of the segment whose endpoints are the foci is called the **center** of the hyperbola, and those three points all fall on a segment called the **transverse axis,** whose endpoints (located on the graph) are called the **vertices**. The **conjugate axis** is perpendicular to the transverse axis at the center of the hyperbola.

Notice in Figures 16.6 and 16.7 that those axes are the length and width of a dotted rectangle drawn on each graph. The diagonals of that rectangle (once extended to the edges of the graph) are the asymptotes of the hyperbola.

Freshly Popped Hyperbola Parts

All the important distances on a hyperbola are labeled using letters you're probably sick to death of seeing in conic sections. All the letters you've come to know and love: *h, k, a, b,* and *c.*

In Figure 16.7, I've drawn a generic hyperbola and labeled all of its parts so you can figure out what they all mean:

♦ **The center of the hyperbola is (*h,k*).** This is no big surprise, since (*h,k*) is always the center of a conic section (except in parabolas, where it's the vertex, since parabolas don't have a center).

♦ **The length of the transverse axis is 2*a*.** In other words, you travel either *a* units up and down or *a* units left and right of the center to reach a vertex.

♦ **The length of the conjugate axis is 2*b*.** Remember, the conjugate axis is the axis that *doesn't* intersect the hyperbola.

♦ **The distance from the center to a focus is *c*.** Trusty old letter *c* has represented the distance from (*h,k*) to the focus since way back on the parabola. Notice that *c* > *a* for hyperbolas. The foci are farther away from the center than the vertices (unlike ellipses, in which the foci were closer to the center than the vertices).

> **CAUTION**
>
> **Kelley's Cautions**
>
> The transverse axis is not necessarily longer than the conjugate axis, so don't automatically refer to the longer of the two axes as *a*, like you did with ellipses. Instead, always use *a* when referring to the transverse axis.

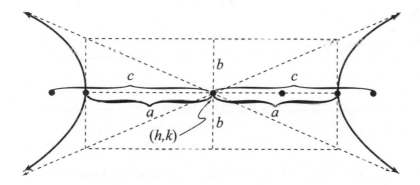

Figure 16.7

A hyperbola centered at (h,k) *with a horizontal transverse axis.*

All of these pieces come together in a standard form that looks very familiar. It looks almost exactly like the standard form of an ellipse, except that it features a negative sign between the fractional terms instead of a positive one.

Standard Form of a Hyperbola

There are two standard forms of a hyperbola, one for horizontal transverse axes and one for transverse axes that are vertical:

$$\frac{(x-b)^2}{a^2} - \frac{(y-k)^2}{b^2} = 1 \qquad \text{or} \qquad \frac{(y-k)^2}{a^2} - \frac{(x-b)^2}{b^2} = 1, \quad \text{if } c = \sqrt{a^2+b^2}$$

Here's the key: The fraction that's positive (and, therefore, comes first in standard form) tells you which way the hyperbola faces. If $(x – h)^2$ appears over the a^2, the hyperbola opens left and right. However, if the $(y – k)^2$ term is first and has that a^2 denominator, the hyperbola opens up and down.

Here's how to remember which is which: Since $(x – h)^2$ contains an x, if it comes first in standard form, the graph must open in the same direction as the x-axis (left and right). On the other hand, if $(y – k)^2$ comes first, the hyperbola must open in the same direction as the y-axis (up and down).

Other than that, you complete the square twice (exactly as you did with ellipses) to convert from general to standard form. Once you've got your hyperbola in standard form, graphing it is not hard; just follow these steps:

1. Graph the center and use a and b to plot the endpoints of the axes, just as you did to graph an ellipse.

2. Draw a rectangle using horizontal and vertical lines whose dimensions match the transverse and conjugate axes.

3. Draw the diagonals of the rectangle you created in Step 2, and extend them to the edges of the coordinate plane.

4. Determine which two of the four endpoints you drew in Step 1 are the vertices. Starting there, draw a parabolic-looking graph that gets close to but never touches the asymptotes.

Even though the two arms of the hyperbola (technically called its *branches*) may look a little like parabolas, they're not—parabolas don't have asymptotes!

Example 3: A hyperbola whose foci are $\left(0, 3 – \sqrt{10}\right)$ and $\left(0, 3 + \sqrt{10}\right)$ has a conjugate axis of length 2. Write the equation of the hyperbola in standard form and graph it.

Kelley's Cautions

Both of the symmetric, nonintersecting branches make up the graph of a single hyperbola. It's the only conic section that consists of two seemingly separate parts, the only one that can't be drawn without lifting your pencil.

Solution: The center of the segment whose endpoints are $\left(0, 3 + \sqrt{10}\right)$ and $\left(0, 3 – \sqrt{10}\right)$ is $(0,3)$, so $h = 0$ and $k = 3$. Since you need to go either $\sqrt{10}$ units up or down from that center to reach one of the foci, $c = \sqrt{10}$. You also know that the conjugate axis (whose length is defined as $2b$) is equal to 2. If $2b = 2$, then $b = 1$. In no time flat, you've got all the constants you need for standard form except for a. Use the formula $c = \sqrt{a^2 + b^2}$ to remedy that:

$$\sqrt{10} = \sqrt{a^2 + 1}$$
$$\left(\sqrt{10}\right)^2 = \left(\sqrt{a^2+1}\right)^2$$
$$9 = a^2$$
$$a = 3$$

How'd You Do That?

You may be asked to find the equations of a hyperbola's asymptotes, but it's a very easy process if you graph the hyperbola first and use the rectangle that's formed by the transverse and conjugate axes.

Each asymptote (remember, every hyperbola has two of them) is guaranteed to pass through the center and exactly two of the corners of the rectangle. To write the equation, pick any two of those three points, and use the technique you reviewed way back in Chapter 2, Example 2a.

It's time to plug all of those constants into standard form, but which one should you use? Since the segment connecting the foci is vertical, the $(y - k)^2$ term should come before the $(x - h)^2$ term:

$$\frac{(y-k)^2}{a^2} - \frac{(x-h)^2}{b^2} = 1$$
$$\frac{(y-3)^2}{3^2} - \frac{(x-0)^2}{1^2} = 1$$
$$\frac{(y-3)^2}{9} - \frac{x^2}{1} = 1$$

To generate the graph of the hyperbola in Figure 16.8, plot the center and then draw the axes. The vertical transverse axis extends 3 units up and down from the center, and the horizontal conjugate axis extends 1 unit left and right of the center. Now draw a rectangle centered at (0,3) whose sides pass through the endpoints of the axes, and extend its diagonals, which will be the asymptotes of the graph.

Figure 16.8

The graph of

$$\frac{(y-3)^2}{9} - \frac{x^2}{1} = 1,\ a\ hyper-$$

bola with a vertical trans-verse axis.

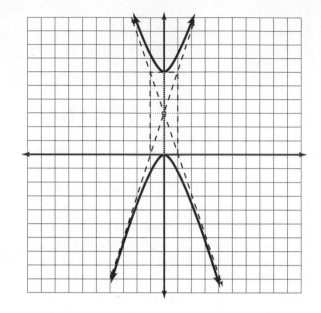

Draw the branches of the graph so that they begin at the endpoints of the transverse axis and stretch toward the asymptotes.

You've Got Problems

Problem 3: Rewrite the equation $25x^2 - 16y^2 + 100x - 32y - 316 = 0$ in standard form and graph it.

The Least You Need to Know

♦ The center of both an ellipse and a hyperbola is (h,k); the distance from the center to one of the foci is c.

♦ In the standard form of an ellipse, $c = \sqrt{a^2 - b^2}$, but in a hyperbola, $c = \sqrt{a^2 + b^2}$.

♦ Converting equations of ellipses and hyperbolas from general form to standard form requires you to complete the square twice, once for x and once for y.

Part 5

Matrices and a Mathematical Montage

In this part, you'll explore the oddly regimented and rectangular world of matrices. It may feel strange that after hundreds of pages containing variables of every size, shape, and ethnic origin, suddenly they're all but gone. It's almost as if some catastrophic event has occurred (perhaps a comet strike, ice age, or gigantic flood), and variables, once the top of the food chain, are all but extinct. Numbers have not only begun to crawl out of their Mesolithic caves, but they are actually organizing and mobilizing in rank and file. Since the book started with a chapter on nothing but numbers, it seems fitting that the last topics are chock full of numbers and even a final exam (don't worry, it's not graded—I even give you the answers).

Chapter 17

Matrix Operations and Calculations

In This Chapter

- ◆ Adding, subtracting, and multiplying matrices
- ◆ Calculating determinants using diagonal multiplication patterns
- ◆ Expanding rows and columns with minors and cofactors
- ◆ Solving systems of equations using Cramer's Rule

There's no way I can possibly discuss matrix operations and not actually bring up the three *Matrix* movies. I have watched all of them numerous times and am somewhat of an expert on the topic. In case you didn't watch any, here's a brief synopsis of each. *The Matrix*: A guy named Neo swallows a red pill, which causes him to learn Kung Fu, wear tight pants, get a gaping metal hole in the back of his head, and meet a cool girl who also wears tight pants. *The Matrix Reloaded:* There's this bus station I don't quite fathom, a guy with a thick French accent you can't understand without subtitles, and Will Smith's wife (who pilots spaceships while wearing sunglasses). *The Matrix Revolutions:* Neo fights in the rain while big machines shoot at other, scarier machines.

Supposedly, there's a whole lot more to these movies, like messianic metaphors, subtly shrouded mythology, and a guy who looks like Colonel Sanders but talks like Stephen Hawking. To make things worse, every *Matrix* movie was harder to understand than the previous one. The basic plot line is simple: Neo = good, Trinity = good, Morpheus = good, Giant Robot Killers That Hurl Bombs and Cut You in Half = bad. Beyond that, it took me multiple viewings to figure out the little details.

The same is true of matrix operations in math. Practically speaking, matrices help you organize and manage huge systems of equations with minimal effort. While they begin quite simply (adding and subtracting matrices could hardly be easier), get a little more tricky ("Whoa, go over multiplying matrices again"), and eventually get downright weird ("Gauss-Jordan elimination? What the heck are you talking about?"). However, the more you practice each one, the easier they are to understand.

If it helps, try wearing tight pants. If nothing else, it might squeeze a little more blood toward your brain to prepare it for these topics, which are unlike any other in precalculus. Without further ado, let's jump in (as Morpheus would say) to see "how deep the rabbit hole goes."

Adding and Subtracting Matrices

A *matrix* is a rectangular block of numbers, written in rows and columns and surrounded by big brackets, like matrix A here:

$$A = \begin{bmatrix} 1 & -3 & 4 & 2 \\ 6 & 0 & 1 & -5 \end{bmatrix}$$

A consists of eight *elements* (or *entries*) organized into two rows and four columns, and thus has the *order* 2×4. (Always write the number of rows before the number of columns when describing the order.) If you so desire, you can write the order of a matrix next to its variable, like this: $A_{2 \times 4}$. If a matrix happens to have the same number of rows and columns, it's called a *square matrix*. When referring to individual elements within matrix A, use the notation a_{ij}, which stands for the element in row i and column j of matrix A. For instance, in the matrix A defined earlier, $a_{12} = -3$ and $a_{24} = -5$.

Talk the Talk

A **matrix** is a rectangular list of numbers (called **elements** or **entries**) organized in rows and columns and surrounded by brackets. The **order** of a matrix describes its dimensions and is written $r \times c$ (where r is the number of rows and c is the number of columns). If a matrix has an equal number of rows and columns, it is called a **square matrix**.

There are three basic matrix operations:

- **Scalar multiplication.** You can multiply a matrix by any number. All you have to do is multiply every single element by that number. (For some reason, math people call the number you're multiplying by a *scalar* rather than a constant.)

$$3\begin{bmatrix} 2 & 9 \\ 0 & -5 \end{bmatrix} = \begin{bmatrix} 3 \cdot 2 & 3 \cdot 9 \\ 3 \cdot 0 & 3(-5) \end{bmatrix} = \begin{bmatrix} 6 & 27 \\ 0 & -15 \end{bmatrix}$$

- **Matrix addition.** You can add two matrices of the same order by adding the corresponding terms.

$$\begin{bmatrix} 9 & 6 \\ -2 & 3 \\ 1 & -5 \end{bmatrix} + \begin{bmatrix} -13 & -7 \\ 5 & -8 \\ 2 & -2 \end{bmatrix} = \begin{bmatrix} 9+(-13) & 6+(-7) \\ -2+5 & 3+(-8) \\ 1+2 & -5+(-2) \end{bmatrix} = \begin{bmatrix} -4 & -1 \\ 3 & 5 \\ 3 & -7 \end{bmatrix}$$

- **Matrix subtraction.** You can subtract one matrix from another by subtracting the corresponding elements.

$$\begin{bmatrix} 4 \\ -6 \\ 3 \end{bmatrix} - \begin{bmatrix} 2 \\ -1 \\ -5 \end{bmatrix} = \begin{bmatrix} 4-2 \\ -6-(-1) \\ 3-(-5) \end{bmatrix} = \begin{bmatrix} 2 \\ -5 \\ 8 \end{bmatrix}$$

Critical Point

Technically, matrix subtraction is a combination of scalar multiplication and addition. When you subtract, you're really just multiplying the second matrix by the scalar -1 and then adding the two matrices.

You've Got Problems

Problem 1: Calculate $-2\begin{bmatrix} 1 & 6 \\ -4 & 3 \end{bmatrix} + 5\begin{bmatrix} -4 & 0 \\ 9 & 7 \end{bmatrix} - 4\begin{bmatrix} -3 & -1 \\ 4 & -6 \end{bmatrix}$.

Multiplying Matrices

Multiplying matrices takes a lot longer than adding or subtracting them because it's not as simple as multiplying corresponding elements. However, before you can even begin to calculate a matrix product, you need to check to see if multiplication is even possible. One requirement must be fulfilled: The product of $A_{m \times n}$ and $B_{p \times r}$ exists only if $n = p$. In other words, the number of columns in the first matrix must match the number of rows in the second matrix.

Kelley's Cautions

Matrix multiplication is not commutative, so you can't assume that $A \cdot B = B \cdot A$ for matrices A and B. Never switch the order of the matrices. If you do, the product may not even exist!

According to this stipulation, the product $(M_{3 \times 4}) \times (N_{3 \times 4})$ does not exist (since M has four columns and N has three rows), but the product $(C_{2 \times 5})(D_{5 \times 1})$ does. (When I say M, N, C, and D, I'm not referring to any matrix in particular here. They're just generic names of matrices with very specific dimensions. For now, it's just the dimensions I want you to focus on.) I know that's weird because matrices C and D are really different, but M and N have the *exact same order*. Unfortunately, no matter how similar M and N are, they'll never multiply with one another. (This is also true of actress Reese Witherspoon and me.)

Let's say you've got two matrices, $A_{m \times n}$ and $B_{n \times p}$, whose product exists (thanks to that matching n dimension). That product will be some matrix C whose order is $m \times p$; it will have the same number of rows as A and the same number of columns as B. Every element c_{ij} in matrix C has to be calculated separately, using the following steps:

1. **Pay attention to row i in matrix A and column j in matrix B.** You'll soon see why those dimensions had to match in such a peculiar way.

2. **Move right across the row and down through the column, multiplying pairs of numbers.** In other words, multiply the leftmost number in the row by the topmost number in the column. Then move one element right in A and one element down in B, and multiply those. You will reach the end of the row and the column simultaneously.

3. **Add the products from Step 2.** The sum of all the products you just calculated will be c_{ij}, the element of C located in the ith row and jth column.

As you multiply across the row and down the column, it helps to mark your place in each matrix with your index fingers. Otherwise, it's really easy to lose your place.

Example 2: If $A = \begin{bmatrix} -3 & 2 \\ 1 & 8 \end{bmatrix}$ and $B = \begin{bmatrix} 7 & 0 & -1 \\ 4 & 3 & -5 \end{bmatrix}$, calculate $A \cdot B$.

Solution: A has two columns, and B has two rows, so $A \cdot B$ exists and will have order 2×3 (since A has two rows and B has three columns). It helps to write the product (which I'll call C) with variables to hold the spots for the elements you will momentarily calculate:

$$\begin{bmatrix} -3 & 2 \\ 1 & 8 \end{bmatrix} \cdot \begin{bmatrix} 7 & 0 & -1 \\ 4 & 3 & -5 \end{bmatrix} = \begin{bmatrix} c_{11} & c_{12} & c_{13} \\ c_{21} & c_{22} & c_{23} \end{bmatrix}$$

To calculate c_{11}, multiply $a_{11} = -3$ times $b_{11} = 7$ and $a_{12} = 2$ times $b_{21} = 4$. Since you've reached the end of A's row and B's column, that's all the multiplication you need to do for element c_{11}—just add those results: $c_{11} = -21 + 8 = -13$. That wasn't so bad, was it? You just do the same thing over and over to fill in the rest of matrix C.

In case you haven't quite understood what I'm saying yet, let me do another element in the product for you. To calculate c_{23}, you multiply the elements in the second row of A (from left to right) by the elements in the third column of B (from top to bottom) and add the results:

$$c_{23} = 1(-1) + 8(-5) = -41$$

Here's what the final product looks like:

$$C = \begin{bmatrix} (-3)7 + 2(4) & (-3)0 + 2(3) & (-3)(-1) + 2(-5) \\ 1(7) + 8(4) & 1(0) + 8(3) & 1(-1) + 8(-5) \end{bmatrix} = \begin{bmatrix} -13 & 6 & -7 \\ 39 & 24 & -41 \end{bmatrix}$$

You've Got Problems
Problem 2: Calculate $\begin{bmatrix} 5 & -3 & -2 \\ -10 & 4 & 8 \\ -1 & 1 & 6 \end{bmatrix} \cdot \begin{bmatrix} 2 & 7 \\ -9 & 3 \\ 0 & -4 \end{bmatrix}$.

Calculating Determinants Using Shortcuts

A *determinant* is a real number that's defined for every square matrix. It has tons and tons of practical uses, but, unfortunately, you won't learn most of them unless you're pursuing a math degree. (I have that backward; in my experience, a math degree pursues you, like in those dreams when you're being chased by a giant monster and, no matter how fast you try to run away, it's always just a few steps behind, snapping its jaws and waving a pocket protector at you.)

To indicate that you're calculating a determinant, use bars around either a matrix's name, $|A|$, or the matrix itself, like this: $\begin{vmatrix} 3 & 4 \\ -1 & 7 \end{vmatrix}$. I've also seen textbooks use the notation det(A), but that's much less common. Even though those vertical lines in the expression $|A|$ may look like absolute value bars, they're not! Bars around numbers indicate an absolute value, but bars around matrices indicate a determinant. Bars around your house indicate that you need to move to a safer neighborhood.

Talk the Talk

Every square matrix has a real number associated with it, called its **determinant**.

Calculating determinants of very small square matrices (2 × 2 and 3 × 3 matrices, specifically) is pretty easy. There are handy shortcuts—specific to each size—that require you only to multiply in a diagonal direction and add or subtract. Once you learn the shortcuts, I'll show you the "real way" (by which I mean "more confusing and work intensive") to calculate determinants.

2 × 2 Matrices

The determinant of the 2 × 2 matrix $\begin{bmatrix} a & b \\ c & d \end{bmatrix}$ is equal to $ad - cb$. Start at the upper-left corner and multiply down diagonally. Then go to the lower-left corner and multiply up diagonally, and subtract that from the first product, as illustrated in Figure 17.1.

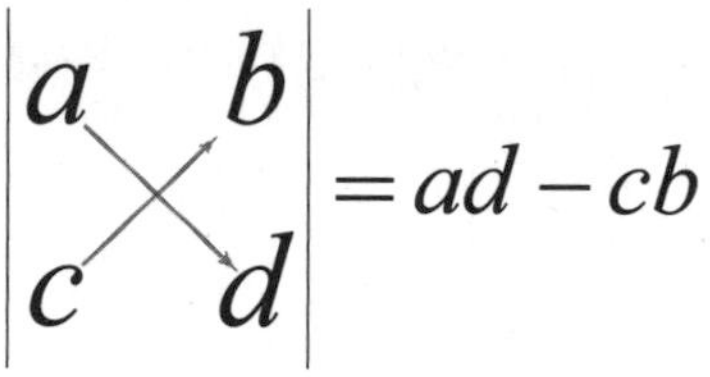

Figure 17.1

Multiply along the arrows and subtract the products to calculate a 2 × 2 determinant.

$$\begin{vmatrix} a & b \\ c & d \end{vmatrix} = ad - cb$$

Critical Point

If a matrix, A, contains only one element, then $|A|$ equals that element.

Example 3: Calculate $\begin{vmatrix} -6 & 3 \\ -2 & -4 \end{vmatrix}$.

Solution: Multiply –6 by the number diagonally across from it (–4), and then subtract from that the number –2 multiplied by its diagonal neighbor (3):

$$\begin{vmatrix} -6 & 3 \\ -2 & -4 \end{vmatrix} = (-6)(-4) - (-2)(3) = 24 + 6 = 30$$

3×3 Matrices

A bit more work is required to find the determinant of a 3×3 matrix. Here are the steps you follow to calculate $\begin{vmatrix} a & b & c \\ d & e & f \\ g & h & i \end{vmatrix}$:

1. **Create a 3×5 matrix by copying columns.** The first three columns match the matrix you're given. The fourth and fifth columns are exact copies of the first and second columns, respectively:

$$\begin{bmatrix} a & b & c & a & b \\ d & e & f & d & e \\ g & h & i & g & h \end{bmatrix}$$

2. **Multiply along six diagonals, adding and subtracting appropriately.** Each diagonal involves three elements. Once again, start in the upper-left corner, and multiply the elements in three downward-pointing diagonals, adding the results. Then scoot down to the lower-left corner and multiply the elements in three upward-pointing diagonals, this time *subtracting* the results, as illustrated in Figure 17.2. The determinant will be $aei + bfg + cdh - gec - hfa - idb$.

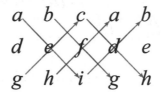

$$= aei + bfg + cdh - gec - hfa - idb$$

Figure 17.2

When calculating a 3×3 determinant, there are three diagonals in each direction. All downward-pointing arrows indicate addition, and upward-pointing arrows indicate subtraction (just like the arrows in a 2×2 matrix).

Example 4: If $A = \begin{bmatrix} 2 & 3 & 0 \\ 4 & 9 & -6 \\ -1 & 1 & 5 \end{bmatrix}$, calculate $|A|$.

Solution: Design a 3×5 matrix so that the three left columns match A, the fourth column matches the first column, and the fifth column matches the second column:

$$\begin{bmatrix} 2 & 3 & 0 & 2 & 3 \\ 4 & 9 & -6 & 4 & 9 \\ -1 & 1 & 5 & -1 & 1 \end{bmatrix}$$

Multiply along the diagonals to get the determinant:

$$|A| = 2(9)(5) + 3(-6)(-1) + 0(4)(1) - (-1)(9)(0) - (1)(-6)(2) - 5(4)(3)$$
$$= 90 + 18 + 0 - 0 + 12 - 60$$
$$= 60$$

You've Got Problems

Problem 4: Calculate $\begin{vmatrix} 3 & -2 & 5 \\ 1 & 9 & 7 \\ 0 & -1 & 4 \end{vmatrix}$.

Calculating Determinants Using Expansions

Just like factoring was a quick, handy, and easy way to solve quadratic equations, it didn't always work. If the quadratic was prime, you had to resort to either using the quadratic formula or completing the square; even though both of those alternatives always worked, they usually took longer and provided more opportunities to screw up, thanks to the more complicated procedures involved.

When it comes to matrices, expansions play the role of trusty, although complex and lengthy, backup to the shortcuts you learned in the preceding section. You can calculate any determinant (no matter what the dimensions) using the expansion method, but you wouldn't want to. Don't get me wrong. It's not all that hard; it just takes longer.

Calculating Minors and Cofactors

The expansion method heavily depends upon two concepts: *minors* and *cofactors*. Both of these things are simply real numbers that correspond to a single element in a matrix. In other words, a matrix containing nine elements has nine minors and cofactors (which are probably different).

Enough generalizations—let's get down to business and be more specific about what these things are. The minor M_{ij} of the matrix element a_{ij} is the determinant of that matrix once the ith row and jth column are eliminated. Therefore, if you start with an $m \times m$ matrix (the dimensions match because it has to be square), the minor will be the determinant of an $(m-1) \times (m-1)$ matrix. Keep in mind that the row and column you eliminate are not *permanently* gone. You just ignore them during the calculation of the minor.

A cofactor is the soul mate of a minor. They have almost everything in common. In fact, the only tangible difference between the two of them is that they sometimes have different moods. If one is positive, the other may be negative; however, they're just as likely to have the same sign. Technically, the cofactor, C_{ij}, of a matrix element a_{ij} is defined like this:

$$C_{ij} = (-1)^{i+j} \cdot M_{ij}$$

Even though that formula looks pretty crazy, it boils down to this: The cofactor of an element is just the minor of that element multiplied by either 1 or –1. Specifically, if $(i+j)$ is an even number, the cofactor and minor will match, but if $(i+j)$ is odd, the cofactor is the opposite of the minor.

Talk the Talk

The **minor** (M_{ij}) of an element a_{ij} is the determinant of the matrix once row i and column j are eliminated. The **cofactor** (C_{ij}) of element a_{ij} is the minor multiplied by $(-1)^{i+j}$.

Example 5: Calculate the minor and cofactor of a_{31} in matrix $A = \begin{bmatrix} 5 & 3 & -3 \\ 2 & 1 & 4 \\ -7 & 0 & 6 \end{bmatrix}$.

Solution: This problem asks you to calculate M_{31} and C_{31}, the minor and cofactor of the element in the third row and first column (–7). To calculate M_{31}, eliminate the third row and first column, and then use the 2 × 2 shortcut to calculate the determinant of what's left:

$$M_{31} = \begin{vmatrix} 3 & -3 \\ 1 & 4 \end{vmatrix} = 12 - (-3) = 15$$

To calculate C_{31}, multiply M_{31} by $(-1)^{3+1}$, which equals $(-1)^4$, or just plain old 1. Therefore, $C_{31} = 15$ as well.

You've Got Problems

Problem 5: Calculate the minor and cofactor for the element -2 in the matrix

$$B = \begin{bmatrix} 1 & -3 & 5 \\ 2 & 9 & -2 \\ 4 & 0 & 1 \end{bmatrix} .$$

Expanding Rows and Columns

Once you've got a good understanding of minors and cofactors, the expansion method (the "long way" of calculating determinants) is a piece of cake. There are only three simple steps to follow:

1. **Choose any single row or column from the matrix.** It doesn't matter which row or column you choose; you'll get the same answer when you're finished.

2. **Multiply each element in the row or column you chose by its cofactor.** This gives you a string of numbers just itching to be combined.

3. **Add those cofactors.** Ah, the itch is scratched.

Kelley's Cautions

Some students get carried away when told they can choose *any* row or column they want to expand and try to expand a diagonal (like they're playing tic-tac-toe), but that won't work.

The act of multiplying the elements of a row or column by their cofactors is called "expanding" the row or column, hence the name of this technique. It's a good idea, when possible, to expand something that contains one or more zeros. Think about it—no matter what its cofactor is, a 0 element times its cofactor always equals 0, so you can save time and skip calculating that cofactor altogether.

Example 6: Given $A = \begin{bmatrix} 3 & 2 & -1 & 4 \\ 8 & 1 & 0 & -6 \\ -5 & 9 & 3 & 5 \\ 2 & 11 & 0 & 8 \end{bmatrix}$, calculate $|A|$.

Solution: There is no shortcut for calculating the determinant of a 4×4 matrix; you have to expand a row or column. I recommend the third column, since it contains two zeros ($a_{23} = 0$ and $a_{43} = 0$).

Multiply each element of the third column by its cofactor:

$$|A| = -1 \cdot C_{13} + 0 \cdot C_{23} + 3 \cdot C_{33} + 0 \cdot C_{43}$$

Thanks to those two zeros, you really don't have to calculate C_{23} and C_{43}. You may not find it necessary, but I like writing out my cofactors as minors before I start the calculations:

$$|A| = -1 \cdot (-1)^{1+3} \cdot M_{13} + 3 \cdot (-1)^{3+3} \cdot M_{33}$$
$$= -M_{13} + 3M_{33}$$

Critical Point

You can check your answer by expanding another row or column to see if you get the same determinant.

It's finally time to calculate the minors. When you temporarily eliminate rows and columns, you'll be left with 3×3 matrices, so use the $3 - 3$ determinant shortcut you learned earlier in the chapter:

$$M_{13} = \begin{vmatrix} 8 & 1 & -6 \\ -5 & 9 & 5 \\ 2 & 11 & 8 \end{vmatrix} = 576 + 10 + 330 - (-108) - 440 - (-40) = 624$$

$$M_{33} = \begin{vmatrix} 3 & 2 & 4 \\ 8 & 1 & -6 \\ 2 & 11 & 8 \end{vmatrix} = 24 + (-24) + 352 \quad 8 \quad (-198) - 128 = 414$$

Now that you know the minors, plug them in:

$$|A| = -M_{13} + 3M_{33} = -624 + 3(414) = 618$$

You've Got Problems

Problem 6: In Problem 4 of this chapter, you were asked to calculate $\begin{vmatrix} 3 & -2 & 5 \\ 1 & 9 & 7 \\ 0 & -1 & 4 \end{vmatrix}$. If you

used the 3×3 shortcut correctly, you got an answer of 132. Verify this answer by redoing the problem with the expansion method.

Cramer's Rule

Do you remember solving systems of equations in algebra? They typically contained two linear equations that looked something like this:

$$\begin{cases} 2x - 6y = 14 \\ -3x + 4y = -16 \end{cases}$$

Your job was to figure out what point (x,y) made both equations true (if such a point existed). Graphically, that meant finding the point where the graphs of those two lines intersected. However, hand-drawn graphs aren't usually accurate enough to get a trustworthy solution.

Critical Point

In rare cases, the equations in a system are actually multiples of one another, resulting in an infinite number of solutions. The two graphs actually overlap at every single point instead of intersecting at just one.

This potential inaccuracy led to two important techniques for solving systems of equations:

♦ **Substitution.** Solve one of the equations for a variable, and plug the result into the other equation.

♦ **Elimination.** Multiply one of the equations by a constant (or both of them by different constants) so that when the equations are added, one of the variables cancels out, allowing you to solve for the other one.

Those techniques work fine, but precalculus offers you a new and fancy, schmantcy way to solve systems of equations, thanks to determinants. The process is called *Cramer's Rule*, and a little matrix manipulation is involved. Whereas calculating minors and cofactors required you to *remove* rows and columns from the matrix, Cramer's Rule requires you to *replace* columns in a matrix with columns from another matrix.

By the way, I understand if you're a little annoyed. After learning all these new and crazy matrix operations, you finally get to use them *to solve problems you knew how to do a long time ago, and it's not even faster!* Sorry about that, but look at the bright side: At least you'll be able to check your answers with the techniques you learned before.

Let me show you how Cramer's Method works in the context of an example:

Example 7: Solve the system: $\begin{cases} 2x - 6y = 14 \\ -3x + 4y = -16 \end{cases}$

Solution: Create a "coefficient matrix" whose elements are the coefficients of x and y in order; I'll call this matrix C:

$$C = \begin{bmatrix} 2 & -6 \\ -3 & 4 \end{bmatrix}$$

Here comes the column replacement I talked about. Create two more matrices, X and Y, that start out as C but have a column replaced by a column containing the constants (14 and –16 from the system of equations). In matrix X, replace the x coefficients (2 and –3) with those constants, and in matrix Y, replace the y coefficients (–6 and 4):

$$X = \begin{bmatrix} 14 & -6 \\ -16 & 4 \end{bmatrix} \qquad Y = \begin{bmatrix} 2 & 14 \\ -3 & -16 \end{bmatrix}$$

Notice that the y coefficients in matrix X are unchanged, as are the x coefficients in matrix Y. Now calculate the determinants of X, Y, and C.

$$|X| = 56 - 96 = -40 \qquad |Y| = -32 + 42 = 10 \qquad |C| = 8 - 18 = -10$$

To get the solution to the system, all you have to do is divide the determinants of X and Y by the determinant of C:

$$x = \frac{|X|}{|C|} \qquad y = \frac{|Y|}{|C|}$$

$$x = \frac{-40}{-10} \qquad y = \frac{10}{-10}$$

$$x = 4 \qquad y = -1$$

The solution to the system is (4,–1).

> **Critical Point**
>
> You can also use Cramer's Rule for systems of three equations with three variables, such as three equations containing x, y, and z. All of the matrices will be 3 × 3, and you'll have to throw in one more matrix, Z, which is generated by replacing the third column of the coefficient matrix by the column of constants.

You've Got Problems

Problem 7: Solve the system: $\begin{cases} -6x + 10y = 2 \\ x + 15y = 1 \end{cases}$

The Least You Need to Know

♦ If $(A_{m \times n})(B_{p \times r})$ exists, then $n = p$, and the product will be an $m \times r$ matrix.

♦ A minor of an element is the determinant of the matrix once the row and columns containing that element are removed; the cofactor is the minor multiplied by either 1 or –1, depending on the location of the element.

♦ You can expand either a row or a column of a matrix to calculate its determinant.

♦ Cramer's Rule allows you to solve systems of equations by calculating determinants of various versions of the system's coefficient matrix.

Chapter 18

The Jagged Little Red Pill of Matrix Applications

In This Chapter

- ◆ Identity and inverse matrices
- ◆ Manipulating matrices with row operations
- ◆ Row-echelon and reduced row-echelon form
- ◆ Solving equations containing matrices

I must display my nerdy tendencies and bring the *Matrix* movies back into the discussion (and to explain the title of this chapter). There's very little overlap between math and pop culture, so when a movie comes along titled after a mathematical concept, well, I just lose control. Forgive me.

Anyway, the central focus of the first *Matrix* movie (also known as "the good one") is that people are hooked up to machines and live their entire lives without realizing that what they think is real and important actually isn't. To escape from the Matrix, our hero, Neo, has to swallow a gigantic red pill given to him by those who want to set him free. Once that pill is ingested, there's no going back.

This chapter (appropriately, the final one in the book containing new topics) is a portal to higher math. Once you ingest the methods and techniques lurking in the pages that follow (either by reading them or simply tearing them out and eating them—go ahead and try it, they're mint flavored), you'll begin to see some of the real power math can harness. (By the way, when I say "mint flavored," I really mean "probably poisonous," so you may want to stick with *reading* the pages instead.)

Miscellaneous Matrix Matters

There are a few basic concepts I need to discuss before we go any further. These will pop up here and there throughout the chapter, so make sure you've got a good handle on each one. They're all related to matrices (surprise, surprise), but every concept is reminiscent of things discussed in the earlier chapters of this book.

Augmented Matrices

When you used Cramer's Rule to solve systems of equations at the end of Chapter 17, specific matrices were involved. You started with a coefficient matrix (featuring all the coefficients from the system) and then replaced its columns one at a time with the constants from the system. After all, the constants are important, so they have to figure into the solution somehow; tediously replacing those columns and calculating all those determinants really is annoying, though, especially when you have more than two equations in the system.

There is another way to rewrite a system as a matrix, one that doesn't require you to swap columns as you go along. If you glue the coefficient matrix and the column of constants together, you'll end up with an *augmented matrix* (literally, a coefficient matrix that's augmented by jamming the constants inside there as well). This system of equations

$$\begin{cases} 4x - 2y = -8 \\ 2x + 5y = 14 \end{cases}$$

translates into this augmented matrix:

$$\left[\begin{array}{cc|c} 4 & -2 & -8 \\ 2 & 5 & 14 \end{array}\right]$$

Notice that a dotted line separates the 2×2 coefficient matrix from the 2×1 column of constants. This boundary clarifies where the coefficients end and the constants begin. An augmented matrix is usually denoted $[A \vdots B]$, where A is the matrix to the left of the dotted line and B is the matrix to the right.

The Identity Matrix

Among the important algebraic properties of Chapter 1, I talked about the identity properties of addition and multiplication. They officially name the identity elements, numbers that won't change a quantity when an operation is applied to it. You've probably always known that adding 0 to something or multiplying something by 1 won't alter its value, but the identity properties harden that mathematical instinct into mathematical fact.

Matrices also have additive and multiplicative identities. The additive identity is just a matrix of matching order chock-full of zeros. In other words, given a matrix $A_{m \times n}$, you can add an $m \times n$ matrix whose elements are all 0, and you'll end up with A. That's not too hard. Actually, it's just the additive identity property applied over and over again, for each element in the matrix.

Kelley's Cautions

Even though matrices have additive and multiplicative identities, the generic term "identity matrix" always means the multiplicative identity matrix (with the diagonal of 1s) unless otherwise noted.

However, the multiplicative identity is slightly more complicated. The *identity matrix* corresponding to any square matrix $B_{m \times m}$ is a matrix of matching order containing all zeros, except in the diagonal beginning in the upper-left corner and stretching down to the lower-right corner. Those elements must all be 1.

Therefore, this 3×3 matrix

$$B = \begin{bmatrix} 3 & -1 & 2 \\ 5 & 7 & -4 \\ -2 & 8 & 0 \end{bmatrix}$$

has an identity matrix, I, that looks like this:

$$I = \begin{bmatrix} 1 & 0 & 0 \\ 0 & 1 & 0 \\ 0 & 0 & 1 \end{bmatrix}$$

How'd You Do That?

In case you're skeptical, here's evidence that the product of B and I is just B. As promised, multiplying by an identity matrix doesn't change a thing (although it seems like a lot of work to not accomplish anything, doesn't it?):

$$B \cdot I = \begin{bmatrix} 3 & -1 & 2 \\ 5 & 7 & -4 \\ -2 & 8 & 0 \end{bmatrix} \cdot \begin{bmatrix} 1 & 0 & 0 \\ 0 & 1 & 0 \\ 0 & 0 & 1 \end{bmatrix}$$

$$= \begin{bmatrix} 3(1)-1(0)+2(0) & 3(0)-1(1)+2(0) & 3(0)-1(0)+2(1) \\ 5(1)+7(0)-4(0) & 5(0)+7(1)-4(0) & 5(0)+7(0)-4(1) \\ -2(1)+8(0)+0(0) & -2(0)+8(1)+0(0) & -2(0)+8(0)+0(1) \end{bmatrix}$$

$$= \begin{bmatrix} 3 & -1 & 2 \\ 5 & 7 & -4 \\ -2 & 8 & 0 \end{bmatrix}$$

Matrix Row Operations

In the last chapter, when discussing Cramer's Rule, I mentioned the elimination technique for solving systems of equations, in which you were allowed to do two interesting things to the equations in any system: multiply an entire equation by any number except 0, and add two equations in a system. You can still do those things for systems represented by a matrix; they're called *row operations*.

Whenever I perform a row operation, I will indicate what I did, what row I did it to (using R_1 to represent Row 1, R_2 to represent Row 2, and so on), and what row contains the result (once again using R_n notation, where n is the row number). Essentially, there are three things you're allowed to do to the rows of a matrix without altering the solution for the system of equations it represents:

♦ **Move rows around.** If you wish the second row were actually the first row, and vice versa, then so be it—just swap their places. Of course, you can't move individual elements; you've got to move the entire row. This is allowed because a system of equations will have the same solution, no matter what order the equations are listed in.

Therefore, the following systems of equations will have the same solution:

$$\begin{bmatrix} 4 & -2 & \vdots & -8 \\ 2 & 5 & \vdots & 14 \end{bmatrix} \quad \text{and} \quad \begin{bmatrix} 2 & 5 & \vdots & 14 \\ 4 & -2 & \vdots & -8 \end{bmatrix}$$

Indicate that row switch by writing either "$R_1 \leftrightarrow R_2$" or "$R_1 \rightarrow R_2$ and $R_2 \rightarrow R_1$."

♦ **Multiply any row by a constant (except for 0).** You did this all the time in the elimination method, and it works well with matrices, too. Let's say you're given the following 3×3 matrix:

Kelley's Cautions

Whenever you apply a row operation to an augmented matrix, the *entire* row (including the left and right sides of the dotted line) is affected.

$$\begin{bmatrix} 3 & -1 & 2 & \vdots & 6 \\ 5 & 7 & -4 & \vdots & -2 \\ -2 & 8 & 0 & \vdots & -5 \end{bmatrix}$$

If you so desire, you can multiply the first row by -5; use the notation $-5R_1 \rightarrow R_1$ to be explicitly clear that all you're doing is multiplying the entire row by -5 and rewriting the row with the results. You'll end up with this matrix:

$$\begin{bmatrix} -5(3) & -5(-1) & -5(2) & \vdots & -5(6) \\ 5 & 7 & -4 & \vdots & -2 \\ -2 & 8 & 0 & \vdots & 5 \end{bmatrix} = \begin{bmatrix} -15 & 5 & -10 & \vdots & -30 \\ 5 & 7 & -4 & \vdots & -2 \\ -2 & 8 & 0 & \vdots & -5 \end{bmatrix}$$

♦ **Add two rows together and replace one of the rows with the result.** As with the previous row operation, this is based in the elimination method. Even though it may feel weird to completely replace a row with the sum of it and another row, it's perfectly valid. Consider the following augmented matrix:

$$\begin{bmatrix} 1 & 6 & \vdots & -11 \\ -5 & -4 & \vdots & 3 \end{bmatrix}$$

Add the two rows and write the result in the second row ($R_1 + R_2 \rightarrow R_2$):

$$\begin{bmatrix} 1 & 6 & \vdots & -11 \\ 1+(-5) & 6+(-4) & \vdots & -11+3 \end{bmatrix} = \begin{bmatrix} 1 & 6 & \vdots & -11 \\ -4 & 2 & \vdots & -8 \end{bmatrix}$$

The second and third row operations are usually used simultaneously. Consider this augmented matrix:

$$\begin{bmatrix} 1 & 6 & \vdots & -11 \\ -5 & -4 & \vdots & 3 \end{bmatrix}$$

Critical Point

The purpose of the operation $5R_1 + R_2 \rightarrow R_2$ is to change the element a_{21} to 0. You may not know why that's important now, but it's a key step in rewriting the matrix in row-echelon form.

Apply the row operation $5R_1 + R_2 \rightarrow R_2$, meaning you should multiply the first row by 5, add that to the second row, and replace the second row with the result. Here's the tricky part: You don't actually want the first row to change at all, even though you're multiplying it by 5 to rewrite the second row. Therefore, you should do all of the multiplying and adding work at the same time inside the second row:

$$\begin{bmatrix} 1 & 6 & | & -11 \\ 5(1)+(-5) & 5(6)+(-4) & | & 5(-11)+3 \end{bmatrix} = \begin{bmatrix} 1 & 6 & | & -11 \\ 0 & 26 & | & -52 \end{bmatrix}$$

You've Got Problems

Problem 1: Perform the following row operations:

a. $-6R_2 \rightarrow R_2$ for $A = \begin{bmatrix} -2 & -3 & 1 & | & 0 \\ 5 & 9 & 2 & | & -6 \\ 4 & -1 & -1 & | & 2 \end{bmatrix}$

b. $-3R_1 + R_2 \rightarrow R_2$ for $B = \begin{bmatrix} 1 & -8 & | & 6 \\ 3 & -5 & | & -2 \end{bmatrix}$

Row-Echelon Form

It's time to roll up your sleeves, strap on your thinking cap, fasten on your bowtie of understanding, and pull up your slacks of intuition. Here comes the crux of the entire chapter: rewriting matrices in two different reduced forms. Essentially, you shake up a matrix using row operations until it meets the requirements of the form you're after.

The first of these two forms is *row-echelon form (the process is called Gaussian elimination)*, and it requires that a matrix meet the following conditions:

♦ **The elements in the diagonal of the matrix must all be 1.** By this, I mean the diagonal from the upper-left corner of the matrix to the lower-right corner, containing elements a_{11}, a_{22}, a_{33}, and so on.

♦ **All of the elements in the columns below the diagonal must be 0.** Once a matrix is in row-echelon form, it looks as though the diagonal of 1s slices the matrix from northwest to southeast, leaving numbers above the diagonal and zeros below it.

◆ **If a row contains nothing but 0 elements, put it at the bottom.** You won't
have to do this very often. It's pretty rare.

The best way to learn this procedure is just
to dive right in and try it, so follow along
with me in the next example. I'm going to
take an augmented 3×4 matrix and put the
coefficient matrix (the 3×3 part of it left of
the dotted line) in row-echelon form. Of
course, any row operations I choose will also
affect things to the left of the dotted line.

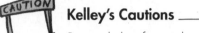

Kelley's Cautions

Row-echelon form takes
a lot of practice! Do enough
examples from your textbook that
the process feels natural and
you're not constantly asking yourself, "What step comes next?"

Example 1: Rewrite the matrix $A = \begin{bmatrix} 2 & 6 & -4 & \vdots & -16 \\ -3 & -1 & 2 & \vdots & -2 \\ 4 & 0 & 10 & \vdots & 13 \end{bmatrix}$ in row-echelon form.

Solution: Your first goal is to get element a_{11} to equal 1. Right now it's 2, so multiply
the entire row by $\frac{1}{2}$ $\left(\frac{1}{2} R_1 \rightarrow R_1 \right)$:

$$\begin{bmatrix} \frac{1}{2}(2) & \frac{1}{2}(6) & \frac{1}{2}(-4) & \vdots & \frac{1}{2}(-16) \\ -3 & -1 & 2 & \vdots & -2 \\ 4 & 0 & 10 & \vdots & 13 \end{bmatrix} = \begin{bmatrix} 1 & 3 & -2 & \vdots & -8 \\ -3 & -1 & 2 & \vdots & -2 \\ 4 & 0 & 10 & \vdots & 13 \end{bmatrix}$$

How'd You Do That?

If you need help with the topics in this chapter, and even if you don't, check out the
free "Linear Algebra Toolkit" website (www.math.odu.edu/~bogacki/lat/), written by
Przemyslaw Bogacki, a professor at Old Dominion University in Virginia.

You can enter any matrix (the interface is very simple), and it will not only do just about
anything you want—such as calculate determinants by expansion, find inverse matrices,
and put matrices in both reduced and regular row-echelon forms—but it will also *show
you how to do every step*.

Now that $a_{11} = 1$, use it to change a_{21} into 0. Multiply the entire row R_1 by the opposite of element a_{21}, and add the result to the second row ($3R_1 + R_2 \rightarrow R_2$):

$$\begin{bmatrix} 1 & 3 & -2 & -8 \\ 3(1)+(-3) & 3(3)+(-1) & 3(-2)+2 & 3(-8)+(-2) \\ 4 & 0 & 10 & 13 \end{bmatrix} = \begin{bmatrix} 1 & 3 & -2 & -8 \\ 0 & 8 & -4 & -26 \\ 4 & 0 & 10 & 13 \end{bmatrix}$$

Continue down the first column, changing the next element (a_{31}) to 0 via the same method: multiply R_1 by the opposite of a_{31}, and then add the result to the third row ($-4R_1 + R_3 \rightarrow R_3$):

$$\begin{bmatrix} 1 & 3 & -2 & -8 \\ 0 & 8 & -4 & -26 \\ -4(1)+4 & -4(3)+0 & -4(-2)+10 & -4(-8)+13 \end{bmatrix} = \begin{bmatrix} 1 & 3 & -2 & -8 \\ 0 & 8 & -4 & -26 \\ 0 & -12 & 18 & 45 \end{bmatrix}$$

Now the first column is taken care of—it has a 1 in the correct diagonal spot and 0s underneath—so move on to the second column. Again, your first goal is to get a 1 in the diagonal, which means $a_{22} = 8$ needs to become 1. Multiply the row by the reciprocal of a_{22} $\left(\frac{1}{8} R_2 \rightarrow R_2 \right)$:

$$\begin{bmatrix} 1 & 3 & -2 & -8 \\ \frac{1}{8}(0) & \frac{1}{8}(8) & \frac{1}{8}(-4) & \frac{1}{8}(-26) \\ 0 & -12 & 18 & 45 \end{bmatrix} = \begin{bmatrix} 1 & 3 & -2 & -8 \\ 0 & 1 & -\frac{1}{2} & -\frac{13}{4} \\ 0 & -12 & 18 & 45 \end{bmatrix}$$

Now that the 1 is where it belongs in the second column, use it to change the element below it ($a_{32} = -12$) into 0, via the row operation $12R_2 + R_3 \rightarrow R_3$:

$$\begin{bmatrix} 1 & 3 & -2 & -8 \\ 0 & 1 & -\frac{1}{2} & -\frac{13}{4} \\ 12(0)+0 & 12(1)+(-12) & 12\left(-\frac{1}{2}\right)+18 & 12\left(\frac{-13}{4}\right)+45 \end{bmatrix} = \begin{bmatrix} 1 & 3 & -2 & -8 \\ 0 & 1 & -\frac{1}{2} & -\frac{13}{4} \\ 0 & 0 & 12 & 6 \end{bmatrix}$$

Critical Point

The steps for row-echelon form follow a definite pattern. Starting with the leftmost column, you change the appropriate diagonal element to 1, use it to change the elements below it to 0, and then move one column left and start over. You're finished when the diagonal is complete.

Now that the second column is finished (it has a 1 in the diagonal and 0 underneath), move to the last column. Change $a_{33} = 12$ into 1 to complete the diagonal $\left(\frac{1}{12}R_3 \to R_3\right)$:

$$
\begin{bmatrix}
1 & 3 & -2 & -8 \\
0 & 1 & -\frac{1}{2} & -\frac{13}{4} \\
\frac{1}{12}(0) & \frac{1}{12}(0) & \frac{1}{12}(12) & \frac{1}{12}(6)
\end{bmatrix}
=
\begin{bmatrix}
1 & 3 & -2 & -8 \\
0 & 1 & -\frac{1}{2} & -\frac{13}{4} \\
0 & 0 & 1 & \frac{1}{2}
\end{bmatrix}
$$

There are no elements beneath a_{33} that must be changed to 0—you've reached the end of the diagonal, the lower-right corner of the coefficient matrix—so you're finished.

You've Got Problems

Problem 2: Rewrite the matrix $\begin{bmatrix} -4 & 3 & -3 \\ 2 & -1 & 1 \end{bmatrix}$ in row-echelon form.

Reduced Row-Echelon Form

Most foods have a "low-something" version these days—there's whole milk or low-fat milk, high-carb or low-carb cookies, and regular Coke or Diet Coke or noncaffeinated Coke or noncaffeinated Diet Coke or lime-flavored Coke or lemon-flavored Coke or diet half-caf vanilla latte with extra espresso and lots of foam but no whipped cream and hold-the-onions Coke. Just one version of a product isn't good enough anymore, so why not embrace a "low-constant" version of the row-echelon form?

As its name suggests, putting a matrix into *reduced row-echelon* form (also called Gauss-Jordan elimination) goes a few steps further with the result of row-echelon form, creating a matrix that contains even more zeros than it did originally. Here's the major difference between the forms: Whereas all the elements of a matrix below the diagonal of 1s in row-echelon form are 0, all of the elements *above* the diagonal of 1s are also 0 in reduced row-echelon form.

Talk the Talk

A matrix in **row-echelon form** has a diagonal containing all 1s, beginning at its upper-left corner, and all the elements below that diagonal must be 0s. If the elements above the diagonal are also 0s, the matrix is in **reduced row-echelon form**.

In row-echelon form, you started in the upper-left corner, forced the diagonal elements into 1s, and then used those 1s to change the elements below into 0s. Reduced row echelon form begins right where row echelon form ends. You start in the lower-right corner (at the other end of the diagonal) and use the 1s to change the elements *above* into 0s. Basically the same process as before, just in a different direction.

The following example picks up where its predecessor (Example 1) left off. It morphs the matrix (already in row-echelon form) into reduced row-echelon form.

Example 2: Rewrite the matrix in reduced row-echelon form:

$$\begin{bmatrix} 1 & 3 & -2 & -8 \\ 0 & 1 & -\frac{1}{2} & -\frac{13}{4} \\ 0 & 0 & 1 & \frac{1}{2} \end{bmatrix}$$

Solution: Start with the 1 in the lower-right corner. Your first goal is to eliminate the element directly above it, $a_{23} = -\frac{1}{2}$. To accomplish this, multiply R_3 by the opposite of a_{23}, and replace R_2 with the sum of R_2 and $R_3\left(\frac{1}{2}R_3 + R_2 \rightarrow R_2\right)$:

$$\begin{bmatrix} 1 & 3 & -2 & -8 \\ \frac{1}{2}(0)+0 & \frac{1}{2}(0)+1 & \frac{1}{2}(1)-\frac{1}{2} & \frac{1}{2}\left(\frac{1}{2}\right)-\frac{13}{4} \\ 0 & 0 & 1 & \frac{1}{2} \end{bmatrix} = \begin{bmatrix} 1 & 3 & -2 & -8 \\ 0 & 1 & 0 & -3 \\ 0 & 0 & 1 & \frac{1}{2} \end{bmatrix}$$

Critical Point

If you're itching for a 3 × 4 matrix to practice reducing into row-echelon and reduced row-echelon form, head to Chapter 19. You'll find more matrix practice problems there (and practice problems for every other chapter as well).

Change $a_{13} = -2$ into 0 using a similar row operation, $2R_3 + R_1 \rightarrow R_1$):

$$\begin{bmatrix} 2(0)+1 & 2(0)+3 & 2(1)+(-2) & 2\left(\frac{1}{2}\right)+(-8) \\ 0 & 1 & 0 & -3 \\ 0 & 0 & 1 & \frac{1}{2} \end{bmatrix} = \begin{bmatrix} 1 & 3 & 0 & -7 \\ 0 & 1 & 0 & -3 \\ 0 & 0 & 1 & \frac{1}{2} \end{bmatrix}$$

Now that all the constants in the third column are correct, move left to the second column. It needs a 0 in the a_{12} spot, above the diagonal. Multiplying R_2 by the opposite of a_{12} and adding the first two rows should do the trick ($-3R_2 + R_1 \rightarrow R_1$):

$$\begin{bmatrix} -3(0)+1 & -3(1)+3 & -3(0)+0 & -3(-3)+(-7) \\ 0 & 1 & 0 & -3 \\ 0 & 0 & 1 & \frac{1}{2} \end{bmatrix} = \begin{bmatrix} 1 & 0 & 0 & 2 \\ 0 & 1 & 0 & -3 \\ 0 & 0 & 1 & \frac{1}{2} \end{bmatrix}$$

You've Got Problems

Problem 3: By the end of Problem 2, you've arrived at a matrix in row-echelon form: $\begin{bmatrix} 1 & -\frac{3}{4} & \frac{3}{4} \\ 0 & 1 & -1 \end{bmatrix}$; put that matrix in reduced row-echelon form.

Why Row-Echelon Forms Even Exist

You may be wondering why row-echelon and reduced row-echelon forms of matrices are even remotely useful (and if you're not, I'm wondering why you're not wondering why they might not be). The answer is actually simple, but before I tell you what it is, please solve this system of equations:

$$\begin{cases} 2x + 6y - 4z = -16 \\ -3x - y + 2z = -2 \\ 4x + 10z = 13 \end{cases}$$

Looks like a lot of work, doesn't it? I'm not going to lie to you; it is a ton of work. After lots of substitutions, the problem barely gets any easier, and if you're anything like me, I'd rather gouge out my eyes than actually try to finish it. Luckily, eye violence

CAUTION

Kelley's Cautions

This system contains three variables, unlike most of the systems you solved in algebra. Therefore, the solution must contain all three variables and is usually written (x,y,z).

is not necessary (it rarely is). You've actually solved this exact system already; in fact, you've nearly solved it twice.

Take all the numbers from that system and create a 3×4 augmented matrix, consisting of a 3×3 coefficient matrix (xs in column 1, ys in column 2, and zs in column 3) on the left and a 3×1 constant matrix on the right, separated by a dotted line:

$$\begin{bmatrix} 2 & 6 & -4 & -16 \\ -3 & -1 & 2 & -2 \\ 4 & 0 & 10 & 13 \end{bmatrix}$$

Look familiar? This is the matrix from Examples 1 and 2 in this chapter. By the end of Example 1, you had this bad boy in row-echelon form:

$$\begin{bmatrix} 1 & 3 & -2 & -8 \\ 0 & 1 & -\frac{1}{2} & -\frac{13}{4} \\ 0 & 0 & 1 & \frac{1}{2} \end{bmatrix}$$

This matrix gives you the coefficients to an entirely new system of equations (with x, y, and z coefficients still in the first, second, and third columns, respectively):

$$\begin{cases} x + 3y - 2z = -8 \\ y - \frac{1}{2}z = -\frac{13}{4} \\ z = \frac{1}{2} \end{cases}$$

This is much easier to solve than the original version of the system. You already know what z equals! All that's left is to find x and y using substitution. Start by plugging z into the second equation in the new system:

$$y - \frac{1}{2}\left(\frac{1}{2}\right) = -\frac{13}{4}$$

$$y = \frac{1}{4} - \frac{13}{4}$$

$$y = -3$$

Now plug y and z into the first equation of the new system:

$$x + 3(-3) - 2\left(\frac{1}{2}\right) = -8$$
$$x = 9 + 1 - 8$$
$$x = 2$$

So, the solution to the system is $\left(2, -3, \frac{1}{2}\right)$. Sure there's a bit of work involved, with the time spent reaching row-echelon form and then substituting into two different equations, but it's better than not solving the system at all, or perhaps solving it and losing an eye. You know what would be even better? Not having to substitute at all.

In Example 2, you finished what Example 1 had started and put the original system of equations in reduced row-echelon form. In case you forgot, this is what you ended up with:

$$\begin{bmatrix} 1 & 0 & 0 & \vdots & 2 \\ 0 & 1 & 0 & \vdots & -3 \\ 0 & 0 & 1 & \vdots & \frac{1}{2} \end{bmatrix}$$

The answer to the system is sitting right there, staring you in the face. As my grandmother used to say, if it were a snake, it would have bitten you by now. (My grandmother always explained things in terms of reptiles, but, hey, that's why we loved her.)

Just read the solution from top to bottom in the rightmost column: $\left(2, -3, \frac{1}{2}\right)$. Alas, there is a reason for row-echelon form, and an even better reason for reduced row-echelon form: They allow you to solve any system of equations using simple row operations.

Inverse Matrices

Nothing is forever, so not even the mighty matrix, sturdy in its rectangular construction and tough, steel-reinforced bracket walls, is completely invulnerable. Just as a number added to its opposite vanishes (with a sum of 0, the additive identity) and a number multiplied by its reciprocal cancels out (turning into 1, the multiplicative identity), a matrix multiplied by its *inverse matrix* will disappear in a cloud of smoke (becoming an identity matrix).

Talk the Talk

When matrix A is multiplied by its **inverse matrix**, A^{-1}, the result is the identity matrix. Note that the notation for an inverse matrix uses the small, elevated $^{-1}$ you also used to denote the inverse of a function, $f^{-1}(x)$.

Here are the steps you should use to find the inverse matrix, A^{-1}, of any matrix A:

1. **Augment the matrix with an identity matrix of matching order.** For instance, if you're finding the inverse of a 2×2 matrix, sew a 2×2 identity matrix onto its right side, using a dotted line to separate the two. Just make sure the dimensions of the two matrices match.

2. **Put the left (original) matrix in reduced row-echelon form.** Make sure you apply all the row operations to the part of the matrix right of the dotted line as well.

3. **Once the left matrix is the identity matrix, the right matrix is the inverse.** The two matrices morph as you apply reduced row-echelon form. What was once the original matrix becomes an identity matrix, and what was once the identity matrix becomes the inverse.

Not all matrices have inverses (those that don't are said to be *singular*), but if an inverse exists, this procedure will hunt it down.

Example 3: If $A = \begin{bmatrix} 4 & 2 \\ -3 & -1 \end{bmatrix}$, calculate A^{-1}.

Solution: Create augmented matrix $[A \vdots I]$, where I is the 2×2 identity matrix:

$$[A \vdots I] = \begin{bmatrix} 4 & 2 & \vdots & 1 & 0 \\ -3 & -1 & \vdots & 0 & 1 \end{bmatrix}$$

Now put A in reduced row-echelon form (so that it looks exactly like I does right now). Your first two steps should be $\frac{1}{4}R_1 \to R_1$ and $3R_1 + R_2 \to R_2$. I'll do those steps simultaneously to save space:

$$\begin{bmatrix} \frac{1}{4}(4) & \frac{1}{4}(2) & \vdots & \frac{1}{4}(1) & \frac{1}{4}(0) \\ 3(1)+(-3) & 3\left(\frac{1}{2}\right)+(-1) & \vdots & 3\left(\frac{1}{4}\right)+0 & 3(0)+1 \end{bmatrix} = \begin{bmatrix} 1 & \frac{1}{2} & \vdots & \frac{1}{4} & 0 \\ 0 & \frac{1}{2} & \vdots & \frac{3}{4} & 1 \end{bmatrix}$$

Apply row operations $2R_2 \to R_2$ and $-\frac{1}{2}R_2 + R_1 \to R_1$ to reach reduced row-echelon form for matrix A:

$$\begin{bmatrix} -\frac{1}{2}(0)+1 & -\frac{1}{2}(1)+\frac{1}{2} & \vdots & -\frac{1}{2}\left(\frac{3}{2}\right)+\frac{1}{4} & -\frac{1}{2}(2)+0 \\ 2(0) & 2\left(\frac{1}{2}\right) & \vdots & 2\left(\frac{3}{4}\right) & 2(1) \end{bmatrix} = \begin{bmatrix} 1 & 0 & \vdots & -\frac{1}{2} & -1 \\ 0 & 1 & \vdots & \frac{3}{2} & 2 \end{bmatrix}$$

The 2 × 2 matrix on the left is the inverse matrix:

$$A^{-1} = \begin{bmatrix} -\dfrac{1}{2} & -1 \\[2mm] \dfrac{3}{2} & 2 \end{bmatrix}$$

If A and A^{-1} truly are inverses, then $(A)(A^{-1})$ should equal the identity matrix. You might as well check it to see if it works:

$$\begin{bmatrix} 4 & 2 \\ -3 & -1 \end{bmatrix}\begin{bmatrix} -\dfrac{1}{2} & -1 \\[2mm] \dfrac{3}{2} & 2 \end{bmatrix} = \begin{bmatrix} 4\left(-\dfrac{1}{2}\right)+2\left(\dfrac{3}{2}\right) & 4(-1)+2(2) \\[2mm] -3\left(-\dfrac{1}{2}\right)+(-1)\left(\dfrac{3}{2}\right) & (-3)(-1)+(-1)(2) \end{bmatrix} = \begin{bmatrix} 1 & 0 \\ 0 & 1 \end{bmatrix}$$

Looks like we got the problem right! Huzzah!

You've Got Problems

Problem 4: Calculate the inverse matrix of $\begin{bmatrix} -\dfrac{1}{3} & 0 \\[2mm] \dfrac{1}{2} & \dfrac{1}{2} \end{bmatrix}$.

Solving Matrix Equations

If you thought using matrices to solve systems of equations was weird, then this will blow your mind: You may run across entire equations written as matrices, like this one:

$$\begin{bmatrix} -1 & 3 \\ -1 & 2 \end{bmatrix}\begin{bmatrix} x \\ y \end{bmatrix} = \begin{bmatrix} 7 \\ -4 \end{bmatrix}$$

Even though that looks very strange (perhaps I'm the only one who thinks it looks a little like the numbers are grazing in three fenced-in pastures, but I digress), you'll solve it a lot like you'd solve much simpler equations.

You'd hardly bat an eyelash if I asked you to solve this equation:

$$\frac{1}{4}x = 3$$

Your instincts should tell you to multiply both sides by 4, but why? You're trying to isolate the variable by using a multiplicative inverse. Any number multiplied by its reciprocal (in this case, 4 is the reciprocal of $\frac{1}{4}$), will vanish, leaving you with $x = 12$.

Apply the same principles to matrix equations. You still want to isolate the variable piece (which is a 2×1 matrix containing x and y in the equation I gave you), but instead of multiplying by the reciprocal of the thing in front of it, multiply both sides of the equation by the one thing that has the power to cancel out a matrix, its inverse matrix.

I'll save you some work and tell you that the inverse matrix of $\begin{bmatrix} -1 & 3 \\ -1 & 2 \end{bmatrix}$ is $\begin{bmatrix} 2 & -3 \\ 1 & -1 \end{bmatrix}$

(use the technique you learned a few pages ago to verify that, if you don't believe me), so multiply it on both sides of the equation:

$$\begin{bmatrix} 2 & -3 \\ 1 & -1 \end{bmatrix} \cdot \begin{bmatrix} -1 & 3 \\ -1 & 2 \end{bmatrix} \cdot \begin{bmatrix} x \\ y \end{bmatrix} = \begin{bmatrix} 2 & -3 \\ 1 & -1 \end{bmatrix} \cdot \begin{bmatrix} 7 \\ -4 \end{bmatrix}$$

You already know that the product of the two leftmost matrices will be a 2×2 identity matrix (after all, they are inverses), so find the product of the matrices on the right:

Kelley's Cautions

Be sure to write the inverse matrix to the left of the matrix you're trying to cancel out *and* to the left of the matrix on the other side of the equation.

$$\begin{bmatrix} 1 & 0 \\ 0 & 1 \end{bmatrix} \cdot \begin{bmatrix} x \\ y \end{bmatrix} = \begin{bmatrix} 2(7)+(-3)(-4) \\ 1(7)+(-1)(-4) \end{bmatrix}$$

$$\begin{bmatrix} x \\ y \end{bmatrix} = \begin{bmatrix} 26 \\ 11 \end{bmatrix}$$

Therefore $x = 26$ and $y = 11$.

You've Got Problems

Problem 5: Solve the matrix equation $\begin{bmatrix} -1 & -2 \\ 0 & 2 \end{bmatrix}\begin{bmatrix} x \\ y \end{bmatrix} = \begin{bmatrix} 4 \\ -3 \end{bmatrix}$.

The Least You Need to Know

♦ The three basic row operations are swapping rows, multiplying a row by a constant, and adding rows.

♦ A matrix, A, in row-echelon form has a diagonal of 1s (starting at element a_{11}), and all the elements below that diagonal must be 0s.

♦ A matrix in reduced row-echelon form has the same diagonal of 1s as in row-echelon form, but all of the elements above and below that diagonal are 0s.

♦ You can use augmented matrices and reduced row-echelon form to calculate an inverse matrix.

19

Final Exam

In This Chapter

- ◆ Measuring your understanding of all major precalculus topics
- ◆ Practicing your skills
- ◆ Determining where you need more practice

Nothing helps you understand math like good, old-fashioned practice, and that's the purpose of this chapter. You can use it however you like, but I suggest one of the following three strategies:

1. As you finish reading each chapter, skip back here and work on the practice problems from that chapter.

2. If you're using this book as a refresher for a class you've already taken, complete this test before you start reading the book. Then go back and read the chapters containing the problems you missed. Once you've reviewed those topics, try these problems again.

3. Save this chapter until the end, and use it to see how much you remember of each topic when you haven't seen it for a while.

Because these problems are just meant for practice and are not meant to teach new concepts, only the answers are given at the end of the chapter, without explanation or justification (unlike the problems in the "You've

Got Problems" sidebars throughout the book). However, these practice problems are designed to mirror those examples, so you can always go back and review if you forgot something or need extra review.

Are you ready? There's a lot of practice ahead of you—more than 115 questions, since many of the 74 problems that follow have multiple parts. (But no one said you have to do them all at one sitting.)

Chapter 1

1. Simplify the expression $\sqrt{-\dfrac{8}{9}+\dfrac{44}{9}}$ and identify all the ways you can classify the result.

2. Identify the properties that justify these statements:

    (a) $8 + (-8) = 0$

    (b) $(x + 4) + (y - 3) = (y - 3) + (x + 4)$

    (c) $(2x)y = 2(xy)$

3. Simplify the expression $\left(\dfrac{\left(a^2\right)^{-4}\cdot b^3\cdot b^5}{a^3 b}\right)^2$.

4. Rewrite as a radical expression $x^{2/7}$.

5. Simplify the radical expression: $4\sqrt{108x} - \sqrt{12x}$.

Chapter 2

6. Solve the equation: $\dfrac{2}{3}\left(x - \dfrac{1}{2}\right) - 5x = 2(x+1) - 6$.

7. Write the equation of the line passing through $(2,-7)$ and $(-4,1)$ in standard form.

8. Graph the equation $x - 3y = 4$.

9. Rewrite in interval notation:

    (a) $x < -2$

    (b) $-5 \leq y < 12$

10. Solve the inequality and graph the solution: $2w - 3 \geq 4 - (w + 1)$.

11. Solve the inequalities and write the solutions in interval notation:

    (a) $|x+3| > 2$

    (b) $3|2x-1| \leq 4$

Chapter 3

12. Classify the following polynomials:

    (a) $-2x^4 + 3x^2 - 5x$

    (b) $x^2 - 1$

13. Simplify the expression $3x(x^2 - 2x -1) - x(x - 5)$.

14. Calculate the product and simplify: $(2x - 3)(x^2 + 4x - 6)$.

15. Calculate the quotient: $(x^4 + 2x^2 - 3x + 1) \div (x^2 + 5x - 2)$.

16. Use synthetic division to calculate $(-x^3 + 4x^2 + 11x - 2) \div (x + 4)$.

17. Simplify the complex expressions:

    (a) $(3 - 4i)(7 + i) - 2(5 - 5i)$

    (b) $(4 - 2i) \div (3 + i)$

Chapter 4

18. Find the greatest common factor of 84 and 154.

19. Factor the polynomials completely:

    (a) $6x^2y^3z^5 + 3xy^7z^3 - 15x^4z^2$

    (b) $2x^4 - 12x^3 - 5x + 30$

    (c) $x^9 - 1$

    (d) $28x^2 - 7$

    (e) $x^2 - 8x + 15$

    (f) $y^2 - 3y - 108$

    (g) $6w^2 - w - 12$

Chapter 5

20. Find the least common denominator of the rational expression:

$$\frac{4}{x^2-9}+\frac{3x}{x-3}-\frac{x^2+2x-13}{x^2+5x+6}.$$

21. Simplify the expression: $\dfrac{4}{x^2-9}+\dfrac{3x}{x-3}-\dfrac{x^2+2x-13}{x^2+5x+6}.$

22. Calculate the product and simplify: $\left(\dfrac{x^2-9x+14}{3x^3-6x^4}\right)\left(\dfrac{2x-1}{x^2-2x-35}\right).$

23. Find the solution(s): $\dfrac{3}{3x-1}-2=\dfrac{25}{9x^2+12x-5}.$

24. Solve and graph the solution of the inequalities:

    (a) $x^2 - 15x + 54 < 0$

    (b) $\dfrac{x-2}{x+6} \geq 3$

Chapter 6

25. If $f(x)=\sqrt{x+1}$, $g(x)=\dfrac{x}{x+4}$, and $h(x) = 2x - 3$, evaluate the following functions:

    (a) $(hg)(4)$

    (b) $(g(h(f(8)))$

26. Evaluate $[\![-8.7]\!]$.

27. Graph the function $f(x)=\left|\dfrac{1}{x}\right|-2$ using transformations.

28. If $g(x) = 3x^2 - 2$, when $x \geq 0$, find $g^{-1}(x)$.

29. Determine how many asymptotes the function $b(x)=\dfrac{x^3-4x-6}{x^2-3x-10}$ has, and give the equation of each.

Chapter 7

30. Solve the quadratic equations using the method indicated:

    (a) $2x^2 + 3x = 27$ (by factoring)

    (b) $2x^2 + 16x + 7 = 0$ (by completing the square)

    (c) $5x^2 + 3x = -1$ (by the quadratic formula)

31. Describe the end behavior of $f(x) = 6x^3 - 8x^4$.

32. Apply Descartes' Rule of Signs to the function $g(x) = -x^3 + 3x^2 + 5x - 1$?

33. Identify all possible rational roots of the function: $h(x) = 6x^3 + 4x^2 - x + 2$.

34. Find all roots of the function $d(x) = 3x^4 - 10x^3 - 4x^2 + 13x + 4$.

Chapter 8

35. Determine the value of x in each logarithmic expression:

    (a) $\log_4 64 = x$

    (b) $\log_3 x = -1$

    (c) $\log_x 128 = \dfrac{7}{2}$

36. Graph the function $f(x) = \log_5 (x + 2) - 1$.

37. Determine the value of $\log_6 31$, rounded to three decimal places.

38. Expand the expression using logarithmic properties: $\ln \dfrac{\sqrt[3]{x} \cdot y^2}{3z}$.

39. Use logarithmic properties to rewrite as a single logarithm:
    $\log_2 w - \log_2 y - 3(\log_2 3 + \log_2 x)$.

Chapter 9

40. Sketch the graph of $j(x) = 3^{-x}$.

41. Solve the equations:

    (a) $2^{x+3} - 3 = 5$

    (b) $\log 4 - \log x = -2$

42. A bacterial population experiences exponential growth during a four-day period. Assume that 200 colonies were present at the beginning of the first day, but by the beginning of the third day, the population had grown to 335. What's the population at the end of the four-day period?

Chapter 10

43. Convert 200° into radians.

44. Convert $\dfrac{12\pi}{5}$ into degrees.

45. Find one positive and one negative coterminal angle for $\theta = \dfrac{4\pi}{9}$.

46. Calculate x in the following diagram; round your answer to the nearest thousandths place.

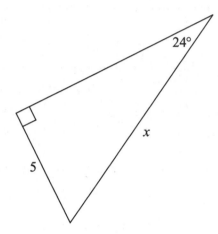

47. Using the unit circle and coterminal angles, but no calculator, evaluate the trigonometric functions:

   (a) $\sin \dfrac{5\pi}{6}$

   (b) $\cos \dfrac{11\pi}{4}$

   (c) $\sin \left(-\dfrac{8\pi}{3} \right)$

Chapter 11

48. Identify the period and amplitude of the following periodic function based on its graph.

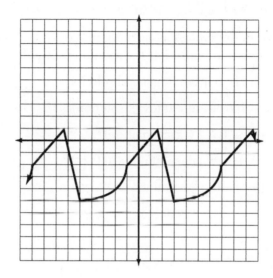

49. Sketch the graph of $f(x) = 2\sin(-x)$ on the interval $[-2\pi, 2\pi]$.

50. Evaluate all six trigonometric functions for the angle $\theta = \dfrac{3\pi}{2}$.

Chapter 12

51. Simplify the expressions:

    (a) $(\sec x)(\csc x) - (\sec(-x))(\csc(-x))$

    (b) $\left(1 + \tan^2\theta\right)\left(1 - \sin^2\theta\right) \cdot \cos\left(\dfrac{\pi}{2} - \theta\right) \cdot \sec\theta$

52. Verify the identities:

    (a) $\sin^2 x - \sin^4 x = (\cos^2 x)(\sin^2 x)$

    (b) $(\csc^2\theta)(\sec^2\theta) = 4\csc^2 2\theta$

    (c) $\sin\left(\theta - \dfrac{\pi}{3}\right) - \sin\left(\theta + \dfrac{\pi}{3}\right) = \sqrt{3}\cos\theta$

Chapter 13

53. Evaluate each, ensuring that your answers are on the correct restricted range:

    (a) $\arcsin\left(-\dfrac{\sqrt{2}}{2}\right)$

    (b) $\operatorname{arccot}\left(-\sqrt{3}\right)$

54. Solve the equations and write the solutions in the form indicated:

    (a) $2(\tan\theta + 1) = 3 + \tan\theta$ (general solution)

    (b) $2\cos^2\theta - 7\cos\theta - 4 = 0$ (exact solution)

    (c) $\cos^2\theta + 4\sin\theta = 3$ (exact solution)

    (d) $1 + \sec\theta = \tan\theta$ (all solutions on $[0,2\pi)$)

    (e) $\sqrt{3}\cot 3\theta - 1 = 0$ (all solutions on $[0,2\pi)$)

Chapter 14

55. Determine the reference angle for $\theta = \dfrac{16\pi}{9}$.

56. If $\tan\theta = -\dfrac{4}{5}$ and $\cos\theta > 0$, evaluate $\sec\theta$.

57. Find the area of a triangle with sides 7, 9, and 14 inches long using Heron's Area Formula, and calculate its largest angle (in degrees) using the Law of Cosines. (Round all answers to the nearest thousandths place.)

58. Given triangle ABC, in which side $a = 5$ is opposite angle $A = 63°$, find the length of the side opposite $B = 50°$.

59. Calculate the area of the following triangle, accurate to the nearest thousandths place.

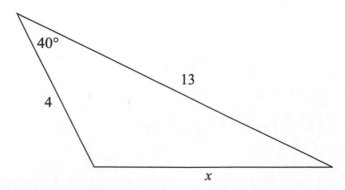

Chapter 15

60. Identify the conic section:

    (a) $4x^2 - 25y^2 - 24x - 50y - 89 = 0$

    (b) $49x^2 + y^2 + 10y - 24 = 0$

    (c) $x^2 + y^2 + 4x - 6y - 23 = 0$

    (d) $2y^2 + x - 16y + 26 = 0$

61. Write the equation of the parabola with focus $(-3,3)$ and directrix $y = -5$ in standard form.

62. Put the parabola $2y^2 + x - 16y + 26 = 0$ in standard form, graph it, and identify the vertex, focus, directrix, and axis of symmetry.

63. Put the circle $x^2 + y^2 + 4x - 6y - 23 = 0$ in standard form, identify its center and radius, and graph it.

Chapter 16

64. Write the equation of the ellipse with a minor axis 8 units long and vertices at points $(-3,-5)$ and $(9,-5)$ in standard form.

65. Put the ellipse $49x^2 + y^2 + 10y - 24 = 0$ in standard form, graph it, and identify its center, foci, vertices, eccentricity, and the lengths of its major and minor axes.

66. Put the hyperbola $4x^2 - 25y^2 - 24x - 50y - 89 = 0$ in standard form and graph it.

Chapter 17

67. Calculate $\begin{bmatrix} 2 & 11 \\ -4 & 6 \\ 9 & 2 \end{bmatrix} - 2 \begin{bmatrix} 5 & -1 \\ 3 & 5 \\ -6 & 1 \end{bmatrix}$

68. Use shortcuts to calculate the determinants:

    (a) $\begin{bmatrix} -12 & 3 \\ 9 & -4 \end{bmatrix}$

    (b) $\begin{bmatrix} 3 & -1 & 4 \\ 2 & 0 & 8 \\ -6 & 3 & -1 \end{bmatrix}$

69. If $A = \begin{bmatrix} -3 & 4 & 1 \\ 9 & -2 & 2 \\ 7 & 5 & -1 \end{bmatrix}$, calculate M_{31} and C_{31}.

70. Calculate $\begin{vmatrix} 1 & -2 & 8 \\ 4 & -9 & -1 \\ 5 & -3 & 0 \end{vmatrix}$ using the expansion method.

71. Solve the system using Cramer's Rule: $\begin{cases} 5x - 2y = 31 \\ 9x + 4y = -5 \end{cases}$

Chapter 18

72. Given the matrix $A = \begin{bmatrix} -2 & 7 & 4 & \vdots & 3 \\ -1 & 8 & 2 & \vdots & -6 \\ -4 & -1 & 5 & \vdots & 2 \end{bmatrix}$, perform the row operation $-3R_3 + R_2 \rightarrow R_2$.

73. Put the matrix $\begin{bmatrix} 1 & -3 & 2 & \vdots & 3 \\ 4 & 1 & -3 & \vdots & -17 \\ -9 & 4 & -1 & \vdots & 12 \end{bmatrix}$ in:

    (a) Row-echelon form.

    (b) Reduced row-echelon form.

74. Solve the equation $\begin{bmatrix} 4 & -2 \\ -1 & 1 \end{bmatrix}\begin{bmatrix} x \\ y \end{bmatrix} = \begin{bmatrix} -6 \\ 1 \end{bmatrix}$ by calculating the inverse of the 2×2 matrix.

Solutions

Chapter 1: (1) 2: natural, whole, integer, rational, real, complex numbers (you could also include positive and even); (2a) additive inverse property; (2b) commutative property because the order of terms changes, not the grouping; (2c) associative property of multiplication; (3) $\dfrac{b^{14}}{a^{22}}$; (4) $\sqrt[7]{x^2}$ or $\left(\sqrt[7]{x}\right)^2$; (5) $22\sqrt{3x}$.

Chapter 2: (6) $x = \dfrac{11}{19}$; (7) $4x + 3y = -13$; (8)

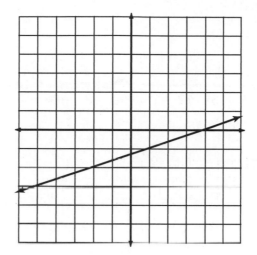

Graph of x − 3y = 4.

(9a) $(-\infty,-2)$; (9b) $[-5,12)$; (10) $[2,\infty)$

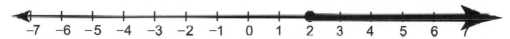

Graph of 2w − 3 ≥ 4 − (w + 1).

(11a) $(-\infty,-5)$ or $(-1,\infty)$; (11b) $\left[-\dfrac{1}{6}, \dfrac{7}{6}\right]$.

Chapter 3: (12a) quartic trinomial; (12b) quadratic binomial; (13) $3x^3 - 7x^2 + 2x$;

(14) $2x^3 + 5x^2 - 24x + 18$; (15) $x^2 - 5x + 29 + \dfrac{-158x + 59}{x^2 + 5x - 2}$; (16) $-x^2 + 8x - 21 + \dfrac{82}{x + 4}$;
(17a) $15 - 15i$; (17b) $1 - i$.

Chapter 4: (18) 14; (19a) $3xz^2(2xy^3z^3 + y^7z - 5x^3)$; (19b) $(2x^3 - 5)(x - 6)$; (19c) $(x - 1)$
$(x^2 + x + 1)(x^6 + x^3 + 1)$. *Hint:* Factor difference of perfect cubes to get $(x^3 - 1)(x^6 + x^3 + 1)$,
and then factor $(x^3 - 1)$ again as the difference of perfect cubes; (19d) $7(2x + 1)(2x - 1)$;
(19e) $(x - 3)(x - 5)$; (19f) $(y - 12)(y + 9)$; (19g) $(3w + 4)(2w - 3)$.

Chapter 5: (20) $(x + 3)(x - 3)(x + 2) = x^3 + 2x^2 - 9x - 18$; (21) $\dfrac{2x^3 + 16x^2 + 41x - 31}{x^3 + 2x^2 - 9x - 18}$;

(22) $-\dfrac{x - 2}{3x^4 + 15x^3}$; (23) $x = 0, -\dfrac{5}{6}$; (24a) $(6,9)$

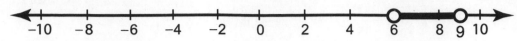

Graph of x² – 15x + 54 < 0.

(24b) [–10,6)

Graph of the solution of $\frac{x-2}{x+6} \geq 3$.

Chapter 6: (25a) $\frac{5}{2}$; (25b) $\frac{3}{7}$; (26) –9; (27)

The graph of $f(x) = \left|\frac{1}{x}\right| - 2$.

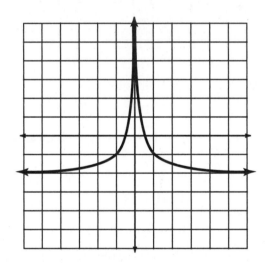

(28) $g^{-1}(x) = \sqrt{\frac{x+2}{3}}$; (29) three asymptotes: $x = -2$, $x = 5$, and $y = x + 3$

Chapter 7: (30a) $x = -\frac{9}{2}, 3$; (30b) $x = -4 - \frac{5\sqrt{2}}{2}, -4 + \frac{5\sqrt{2}}{2}$; (30c) $\frac{-3 - i\sqrt{11}}{10}, \frac{-3 + i\sqrt{11}}{10}$;

(31) the right and left ends of the graph both point down; (32) $g(x)$ has either two or zero positive roots, and $g(x)$ has exactly one negative root; (33) $-2, -1, -\frac{2}{3}, -\frac{1}{2}, -\frac{1}{3}, -\frac{1}{6}$,

$\frac{1}{6}, \frac{1}{3}, \frac{1}{2}, \frac{2}{3}, 1, 2$; (34) -1, $\frac{4}{3}, \frac{3 + \sqrt{13}}{2}, \frac{3 - \sqrt{13}}{2}$

Chapter 8: (35a) 3; (35b) $\frac{1}{3}$; (35c) 4; (36)

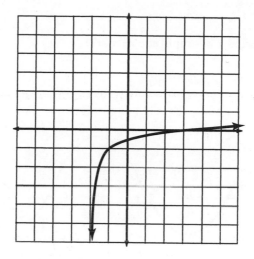

Graph of $f(x) = log_5 (x + 2) - 1.$

(37) 1.917; (38) $\frac{1}{3}\ln x + 2\ln y - \ln 3 - \ln z$; (39) $\log_2 \dfrac{w}{27x^3 y}$

Chapter 9: (40)

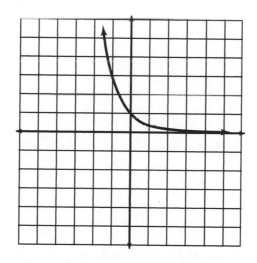

The graph of $j(x) = 3^{-x}.$

(41a) $x = 0$; (41b) $x = 400$; (42) approximately 561 colonies

Chapter 10: (43) $\frac{10\pi}{9}$; (44) 432°; (45) $\frac{22\pi}{9}, -\frac{14\pi}{9}$; (46) $x = 12.293$; (47a) $\frac{1}{2}$;

(47b) $-\frac{\sqrt{2}}{2}$; (47c) $-\frac{\sqrt{3}}{2}$

Chapter 11: (48) period = 8; amplitude = 3; (49)

The graph of f(x) = 2sin(−x).

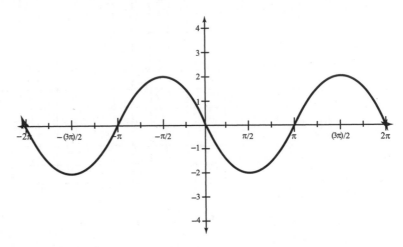

(50) $\cos\frac{3\pi}{2} = 0, \sin\frac{3\pi}{2} = -1, \tan\frac{3\pi}{2}$ is undefined, $\cot\frac{3\pi}{2} = 0, \sec\frac{3\pi}{2}$ is undefined,

$\csc\frac{3\pi}{2} = -1$

Chapter 12: (51a) 2sec x csc x; (51b) tan θ;

(52a)

Factor sin²x from the left side: $\sin^2x(1 - \sin^2x) = (\cos^2x)(\sin^2x)$

Replace $(1 - \sin^2x)$ with cos²x (Pythagorean Identity): $\sin^2x(\cos^2x) = (\cos^2x)(\sin^2x)$

Commutative property of multiplication: $(\cos^2x)(\sin^2x) = (\cos^2x)(\sin^2x)$

(52b)

Apply reciprocal identities: $\dfrac{1}{\sin^2\theta \cdot \cos^2\theta} = \dfrac{4}{\sin^2 2\theta}$

Cross-multiply the proportion: $4\sin^2\theta \cdot \cos^2\theta = \sin^2 2\theta$

Apply double-angle formula: $4\sin^2\theta \cdot \cos^2\theta = (2\sin\theta\cos\theta)^2$

Square right side: $4\sin^2\theta \cdot \cos^2\theta = 4\sin^2\theta \cdot \cos^2\theta$

(52c)   Apply sum and difference formulas: $\left(\sin\theta\cos\dfrac{\pi}{3}+\cos\theta\sin\dfrac{\pi}{3}\right)-\left(\sin\theta\cos\dfrac{\pi}{3}-\cos\theta\sin\dfrac{\pi}{3}\right)=\sqrt{3}\cos\theta$

Evaluate trig functions: $\left(\dfrac{1}{2}\sin\theta+\dfrac{\sqrt{3}}{2}\cos\theta\right)-\left(\dfrac{1}{2}\sin\theta-\dfrac{\sqrt{3}}{2}\cos\theta\right)=\sqrt{3}\cos\theta$

Simplify: $2\left(\dfrac{\sqrt{3}}{2}\cos\theta\right)=\sqrt{3}\cos\theta$

$$\sqrt{3}\cos\theta=\sqrt{3}\cos\theta$$

Chapter 13: (53a) $-\dfrac{\pi}{4}$; (53b) $\dfrac{5\pi}{6}$; (54a) $\theta=\dfrac{\pi}{4}+2k\pi,\dfrac{5\pi}{4}+2k\pi$ where k is an integer

(you could also write $\theta=\dfrac{\pi}{4}+k\pi$); (54b) $\theta=\dfrac{2\pi}{3}$; (54c) $\theta=\arcsin\left(2-\sqrt{2}\right)\approx0.626$

(54d) $\theta=0,\pi$; (54e) $\theta=\dfrac{\pi}{9},\dfrac{4\pi}{9},\dfrac{7\pi}{9},\dfrac{10\pi}{9},\dfrac{13\pi}{9},\dfrac{16\pi}{9}$

Chapter 14: (55) $\dfrac{2\pi}{9}$; (56) $\sec\theta=\dfrac{\sqrt{41}}{5}$; (57) Area = $12\sqrt{5}\approx26.833$, largest angle = 121.588°; (58) $b=4.299$; (59) 16.712

Chapter 15: (60a) hyperbola; (60b) ellipse; (60c) circle; (60d) parabola;

(61) $y=\dfrac{1}{16}(x+3)^2-1$; (62) $x=-2(y-4)^2+6$, vertex = (6,4), focus = $\left(\dfrac{47}{8},4\right)$, directrix: $x=\dfrac{49}{8}$, axis of symmetry: $y=4$

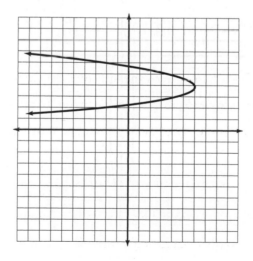

The graph of $2y^2+x-16y+26=0$.

(63) $(x+2)^2+(y-3)^2=36$, center = (−2,3), radius = 6

The graph of $x^2 + y^2 + 4x - 6y - 23 = 0$.

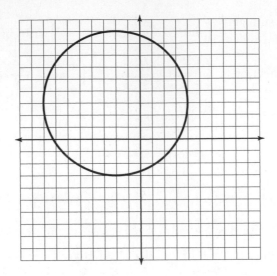

Chapter 16: (64) $\dfrac{(x-3)^2}{36} + \dfrac{(y+5)^2}{16} = 1$; (65) $\dfrac{x^2}{1} + \dfrac{(y+5)^2}{49} = 1$, center = (0,–5), foci = $\left(0, -5 - 4\sqrt{3}\right)$ and $\left(0, -5 + 4\sqrt{3}\right)$, vertices = (0,2) and (0,–12), eccentricity = $\dfrac{4\sqrt{3}}{7} \approx 0.990$, major axis is 14 units long, minor axis is 2 units long

The graph of $49x^2 + y^2 + 10y - 24 = 0$.

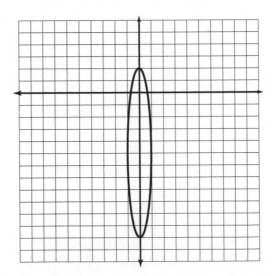

(66) $\dfrac{(x-3)^2}{25} - \dfrac{(y+1)^2}{4} = 1$

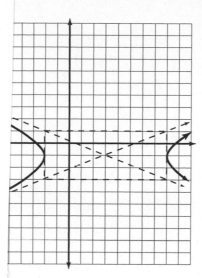

The graph of $4x^2 - 25y^2 - 24x - 50y - 89 = 0$.

$\bigg]$; (68a) 21; (68b) −2; (69) $M_{31} = C_{31} = 10$; (70) 271;

$$\begin{bmatrix} 7 & 4 & \vdots & 3 \\ 11 & -13 & \vdots & -12 \\ -1 & 5 & \vdots & 2 \end{bmatrix} ; (73a) \begin{bmatrix} 1 & -3 & 2 & \vdots & 3 \\ 0 & 1 & -\dfrac{11}{13} & \vdots & -\dfrac{29}{13} \\ 0 & 0 & 1 & \vdots & 5 \end{bmatrix}$$

$A^{-1} = \begin{bmatrix} \dfrac{1}{2} & 1 \\ \dfrac{1}{2} & 2 \end{bmatrix}$, solution = (−2,−1)

A

Solutions to "You've Got Problems" Sidebars

Here are detailed solutions for all the "You've Got Problems" sidebars throughout the book. I suggest that you turn back here to look at the answers only once you've tried your best or are hopelessly stuck; you should find just enough information to get you through any important or tricky steps.

Don't just read the problem and then flip back here to read the answer! If you *do the problem yourself first*, you'll actually master the concept on your own.

Chapter 1

1. Rational, real, and complex numbers. To combine -5 and $\frac{2}{3}$, you first need to get common denominators: $-\frac{15}{3} + \frac{2}{3} = -\frac{13}{3}$. Because the result is negative, it cannot be a natural or whole number. Because it is a fraction, $-\frac{13}{3}$ is not an integer, but is definitely rational. Any rational number is automatically a real number, and any real is automatically complex.

2. (a) Symmetric property. The sides of the equation are reversed.

(b) Commutative property of multiplication. The order of the numbers has changed.

(c) Distributive property. The 4 is multiplied through a set of parentheses.

(d) Inverse property of multiplication. Any number times its reciprocal equals 1.

3. $\dfrac{x^8}{y^6}$. Start inside the parentheses and subtract the powers of the common bases: $(x^{3-7}y^{5-2})^{-2} = (x^{-4}y^3)^{-2}$. Multiply each exponent in the parentheses by -2: x^8y^{-6}. Finally, rewrite the answer without a negative exponent.

4. (a) $\sqrt{y}$. Even though the expression technically equals $\sqrt[2]{y^1}$, both the index 2 and exponent 1 are implied.

(b) $\sqrt{b^3}$ or $\left(\sqrt{b}\right)^3$; both expressions are equal and correct.

5. (a) 1024. Rewrite as a radical expression: $16^{5/2} = \left(\sqrt{16}\right)^5$. Simplify the expression:
$$\left(\sqrt{16}\right)^5 = 4^5 = 1024 .$$

(b) $4|x|\sqrt{7}$. Simplify each radical:
$$\sqrt{4 \cdot 7 \cdot x^2} + \sqrt{9 \cdot 7 \cdot x^2} - \sqrt{7x^2} = 2|x|\sqrt{7} + 3|x|\sqrt{7} - |x|\sqrt{7}$$

Since all the indexes and radicands match, combine the coefficients ($2 + 3 - 1 = 4$).

Chapter 2

1. $x = -6$. Distribute -5: $-\dfrac{10}{3}x - 30 = \dfrac{7}{3}x + 4$. Subtract $\dfrac{7}{3}x$ from and add 30 to both sides of the equation: $-\dfrac{17}{3}x = 34$. Multiply both sides by $-\dfrac{3}{17}$ to isolate x:
$\left(-\dfrac{3}{17}\right)\left(-\dfrac{17}{3}x\right) = \left(-\dfrac{3}{17}\right)\left(\dfrac{34}{1}\right)$. Simplify to get $x = -6$.

2. $y = -6x - 45$. The slope of v is $-\dfrac{A}{B} = -\dfrac{6}{1} = -6$. Since v and w are parallel, their slopes are equal. Use $m = -6$ and $(a,b) = (-7,-3)$ in the point-slope formula, and simplify:
$$y - (-3) = -6\left(x - (-7)\right)$$
$$y + 3 = -6x - 42$$

Solve for y to put this answer in slope-intercept form.

3. Substituting $x = 0$ gives you $-3y = -5$, so $y = \frac{5}{3}$. Setting $y = 0$ results in $x = -5$. Plot the points $\left(0, \frac{5}{3}\right)$ and $(-5,0)$, and connect the dots to get the following graph:

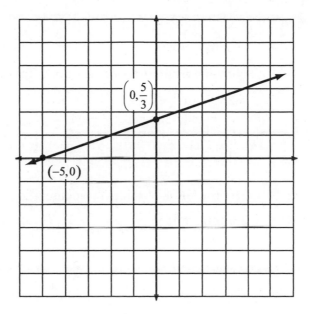

4. (a) $[-1,\infty)$. The lower boundary is finite and included, and the upper boundary is infinite.

(b) $(9,13]$. The upper boundary is included, but the lower is not.

(c) $(-\infty,4)$ or $(4,\infty)$. If x cannot be 4, it must be either less than 4 or greater than 4. Translate those two statements ($x < 4$ or $x > 4$) into intervals.

5. $y \geq -\frac{9}{2}$. Distribute the 9, and then subtract $9y$ from both sides of the inequality: $-4y \leq 18$. Divide both sides by -4, remembering to reverse the inequality sign since you're dividing by a negative number: $y \geq -\frac{9}{2}$. The graph is a number line with a solid dot at $-\frac{9}{2}$ and a dark arrow pointing right from that endpoint.

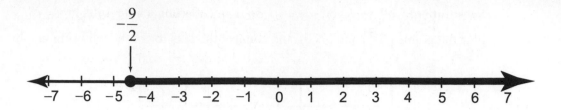

6. (a) $x \geq 7$ or $x \leq 3$. Subtract 1 from each side: $|x - 5| \geq 2$. Then rewrite as two inequalities: $x - 5 \geq 2$ or $x - 5 \leq -2$. Solve each separately.

(b) $-\frac{4}{3} < x < \frac{4}{3}$. Isolate the absolute values by adding 4 to both sides: $|3x| < 4$.

Now rewrite as a compound inequality: $-4 < 3x < 4$. Divide everything by 3 to solve.

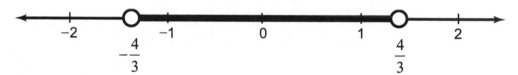

Chapter 3

1. (a) Cubic monomial. There's only one term, and it's raised to the third power.

   (b) Quadratic trinomial. There are three terms, and the highest exponent is 2.

   (c) Linear binomial. The degree of this two-termed polynomial is 1.

2. $-12x^2 - 3x + 5y^2 + 10y - xy$. Distribute $-3x$ and $5y$ through the corresponding sets of parentheses: $-12x^2 + 9xy - 3x - 10xy + 5y^2 + 10y$. The only like terms in the entire expression are $9xy$ and $-10xy$; combine them to get $9xy - 10xy = -xy$, and leave the other terms alone.

3. $-2a^3 - 9a^2b + 43ab^2 - 20b^3$. Distributing first the $-2a$ and then $5b$ through the polynomial on the right gives you this:

$$(-2a)(a^2) + (-2a)(7ab) + (-2a)(-4b^2) + (5b)(a^2) + 5b(7ab) + 5b(-4b^2)$$

$$= -2a^3 - 14a^2b + 8ab^2 + 5a^2b + 35ab^2 - 20b^3$$

Don't forget to add the exponents of common bases when you multiply; that's why $(-2a)(7ab) = -14a^{1+1}b = -14a^2b$. Finally, combine like terms to simplify.

4. $x^2 + 3x - \dfrac{4}{x-3}$. Note that you must add $0x^2$ to the dividend as you set up the problem.

$$
\begin{array}{r}
x^2 + 3x \\
x-3 \overline{\smash{)}\, x^3 + 0x^2 - 9x - 4} \\
\underline{-x^3 + 3x^2} \\
3x^2 - 9x \\
\underline{-3x^2 + 9x} \\
0 - 4
\end{array}
$$

Since you can't multiply x by anything to get -4, you're done. The quotient is $x^2 + 3x$, and the remainder is -4.

5. $x^3 - 4x^2 - 6x - 12 - \dfrac{29}{x-2}$. Remember to write the opposite of the divisor's constant in the box: 2, not -2. Since the dividend has no x term, make sure there's a 0 in the x's spot of the coefficient list:

$$
\begin{array}{r|rrrrr}
2 & 1 & -6 & 2 & 0 & -5 \\
  &   & 2 & -8 & -12 & -24 \\
\hline
  & 1 & -4 & -6 & -12 & -29
\end{array}
$$

6. To calculate $c + d$, combine like terms: $(1 - 6i) + (4 + 3i) = 5 - 3i$. You need to distribute -1 to subtract c and d: $(1 - 6i) - (4 + 3i) = 1 - 6i - 4 - 3i = -3 - 9i$.

Multiply c and d to get $4 + 3i - 24i - 18i^2$. When you combine like terms, don't forget that $i^2 = -1$; you'll get an answer of $22 - 21i$. Here's the work for the quotient $c \div d$:

$$
\frac{1-6i}{4+3i} \cdot \frac{4-3i}{4-3i} = \frac{4-3i-24i+18i^2}{16-12i+12i-9i^2} = \frac{-14-27i}{25} = -\frac{14}{25} - \frac{27}{35}i
$$

Chapter 4

1. $2^2 \cdot 3^2 = 36$. The prime factorization of 72 is $2^3 \cdot 3^2$, and 180 factors into $2^2 \cdot 3^2 \cdot 5$, as you can see by the factor trees that follow.

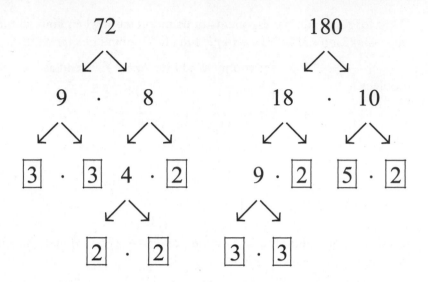

The common factors are 2 and 3, so compare the exponents and use the lesser of the two for each. That means 2^2 instead of 2^3, and don't forget 3^2 (both factorizations contain this, so there's no need to choose which exponent of 3 to use). Multiply together to get the greatest common factor.

2. $-3a^2bc^2(6ab^4 + 3b^2c^5 + 2ac^3)$. The greatest common factor of the coefficients is 3, but since all the terms are negative, make it $-1 \cdot 3 = -3$. Every term contains an a, b, and c, so choose the lowest exponent of each: a^2, b, and c^2. That gives you a greatest common factor of $-3a^2bc^2$. Now write the polynomial in factored form and simplify:

$$-3a^2bc^2\left(\frac{-18}{-3}\,a^{3-2}b^{5-1}c^{2-2} + \frac{-9}{-3}\,a^{2-2}b^{3-1}c^{7-2} + \frac{-6}{-3}\,a^{3-2}b^{1-1}c^{5-2}\right)$$

3. $(2w - 3)(5x - 6y)$. Rewrite the polynomial by pairing its first two and last two terms: $(10xw - 15x) + (-12wy + 18y)$. Factor each pair separately: $5x(2w - 3) + 6y(-2w + 3)$. Since the binomials are opposites, factor a -1 out of the second one to get $5x(2w - 3) - 6y(2w - 3)$. To finish, factor out the common binomial.

4. (a) $3(x^2 + 4y^2)(x + 2y)(x - 2y)$. Begin by pulling out the greatest common factor: $3(x^4 - 16y^4)$. Inside the parentheses you have the difference of perfect squares: $3(x^2 + 4y^2)(x^2 - 4y^2)$. You're not finished—notice that $(x^2 - 4y^2)$ is another difference of perfect squares; so factor it into $(x + 2y)(x - 2y)$.

(b) $(r - 5)(r^2 + 5r + 25)$. You've got a difference of perfect cubes, so use the corresponding formula with $a = r$ and $b = 5$:

$$(a^3 - b^3) = (a - b)(a^2 + ab + b^2)$$

$$(r^3 - 5^3) = (r - 5)(r^2 + (r)(5) + (5)^2)$$

5. (a) $(y + 3)(y + 4)$. The numbers 3 and 4 both add up to 7 and have a product of 12.

(b) $3(x + 7)(x - 4)$. Start by factoring out the greatest common factor of 3: $3(x^2 + 3x - 28)$. Only the numbers 7 and –4 add up to 3 and multiply to give you –28. Be careful to not reverse the signs and use the numbers –7 and 4, because the sum would be incorrect: $-7 + 4 = -3$, not 3.

6. $(x + 2)(3x - 4)$. The only two numbers that have a sum of 2 and a product of –24 are –4 and 6. Rewrite the x-coefficient and distribute the x to get $3x^2 \ 4x + 6x - 8$. Factor by grouping: $x(3x - 4) + 2(3x - 4) = (3x - 4)(x + 2)$.

Chapter 5

1. $(x - 2)(x - 1)^2(x + 5)$. Factor the denominators (using the method from Chapter 4, in the section "Factoring Quadratic Trinomials") to get

$$\frac{x^2}{(x-2)(x-1)} + \frac{x-2}{(x-1)^2} - \frac{5x^2}{(x+5)(x-2)}$$. (Note that $x^2 - 2x + 1 = (x - 1)(x - 1) =$
$(x - 1)^2$ in the second fraction.) The unique factors are $(x - 2)$, $(x - 1)$, and $(x + 5)$; use the greatest power of each to construct the least common denominator—the order of the factors doesn't matter.

2. $\dfrac{28x + 11}{12x + 15}$. The least common denominator of this expression is $3x(4x + 5)$.

   Rewrite each fraction with that denominator:

$$\frac{2}{3} \cdot \frac{x(4x+5)}{x(4x+5)} - \frac{x-1}{3(4x+5)} \cdot \frac{x}{x} + \frac{7x^2}{x(4x+5)} \cdot \frac{3}{3} = \frac{2x(4x+5) - x(x-1) + 3 \cdot 7x^2}{3x(4x+5)}$$

Combine the numerators and simplify to get your final answer.

$$\frac{8x^2 + 10x - x^2 + x + 21x^2}{3x(4x+5)} = \frac{28x^2 + 11x}{3x(4x+5)} = \frac{\cancel{x}(28x+11)}{3\cancel{x}(4x+5)} = \frac{28x+11}{12x+15}$$

3. $\dfrac{4x^3 - 12x^2}{x^2 + 4x + 3}$. Factor and then multiply the fractions:

$$\frac{x(2x-3)}{(x+3)(x+3)} \cdot \frac{4x(x^2-9)}{(2x-3)(x+1)} = \frac{x\cancel{(2x-3)}(4x)\cancel{(x+3)}(x-3)}{\cancel{(x+3)}(x+3)\cancel{(2x-3)}(x+1)}$$

Expand (multiply the factors in) the numerator and denominator to finish.

4. $\dfrac{ab^2 + ab + a - b^2 - b - 1}{a+4}$. Factor the expressions, change the division sign to multiplication, and take the reciprocal of the fraction to the right of the division sign:

$$\frac{a}{b-1} \cdot \frac{2a^2 - a - 1}{a+4} \div \frac{2a^2 + a}{b^3 - 1} = \frac{a}{b-1} \cdot \frac{(2a+1)(a-1)}{a+4} \cdot \frac{(b-1)(b^2+b+1)}{a(2a+1)}$$

Write as one fraction and simplify.

$$\frac{\cancel{a}\,\cancel{(2a+1)}\,(a-1)\,\cancel{(b-1)}\,(b^2+b+1)}{\cancel{(b-1)}\,(a+4)\,\cancel{(a)}\,\cancel{(2a+1)}} = \frac{ab^2 + ab + a - b^2 - b - 1}{a+4}$$

5. $x = \dfrac{31}{2}$. Factor the rightmost denominator to get $5(x-7)$, which also happens to be the common denominator. Therefore, multiply both sides of the equation by $5(x-7)$:

$$\frac{\cancel{5}\,(x-7)\cdot 3}{\cancel{5}} - \frac{5\,\cancel{(x-7)}\cdot 2}{\cancel{x-7}} = \frac{\cancel{5}\,\cancel{(x-7)}\cdot x}{\cancel{5}\,\cancel{(x-7)}}$$

$$3x - 21 - 10 = x$$

$$2x = 31$$

Divide both sides by 2 to finish; when checked, the solution is valid.

6. [0,3]. Factor out the greatest common factor of x to get $x(x-3) \le 0$. Set x and $(x-3)$ equal to 0, and solve to get critical numbers of 0 and 3. Choosing one test value on each interval $(-\infty, 0]$, $[0,3]$, and $[3,\infty)$ reveals that only the values on [0,3] make the inequality true, so it's the solution. Use solid dots on the graph because of the inequality symbol ($\le$).

7. (–8,2). Subtract $\dfrac{1}{5}$ from both sides, and then use the least common denominator of $5(x-2)$ to combine the terms into one fraction: $\dfrac{4x+32}{5x-10} < 0$. Set the numerator and denominator equal to 0 to get critical numbers of $x = -8$ and $x = 2$. Use test values to get the solution: (–8,2). Both critical numbers should be marked on the graph with open dots.

Chapter 6

1. (a) $2w - 5$. Divide $h(w)$ by $k(w)$, and notice that you can simplify the fraction by factoring the numerator:

$$\left(\frac{h}{k}\right)(w) = \frac{2w^2 - 3w - 5}{w + 1} = \frac{(2w - 5)\,(w + 1)}{w + 1} = 2w - 5$$

   (b) -5. The function $(h \circ k)(-1)$ means the same thing as $h(k(-1))$. Since $k(-1) = 0$, plug that into $h(w)$: $h(0) = 2(0)^2 - 3(0) - 5 = -5$.

2. Three transformations are applied to the graph of x^2 to get the graph of $h(x)$: a reflection across the x-axis (which will flip the graph upside-down), a horizontal shift of 3 units left, and a vertical shift of 1 unit down.

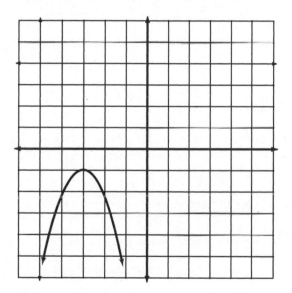

3. $g^{-1}(x) = \frac{1}{2}x + 3$. Rewrite $g(x)$ as y, and then switch the x and y variables to get $x = 2y - 6$. Solve for y: $y = \frac{x + 6}{2}$ (which can be rewritten by splitting the fraction into two parts $\left(y = \frac{x}{2} + \frac{6}{2}\right)$ and simplifying: $y = \frac{1}{2}x + 3$). The final answer should contain $g^{-1}(x)$ instead of y.

4. Two asymptotes: $x = 8$ and $y = 3$. Factor the function first: $g(x) = \frac{(x - 4)(3x + 1)}{(x - 4)(x - 8)}$.

   Since an x-value of 8 makes the denominator 0 but not the numerator, it represents a vertical asymptote: $x = 8$. (Notice that $x = 4$ makes *both* parts of the fraction 0, so it's not an asymptote.) The degrees of the numerator and denominator are equal, so the horizontal asymptote will be the quotient of the leading coefficients $\left(\frac{3}{1}\right)$ and will have the equation $y = 3$.

Chapter 7

1. $x = 1 + \sqrt{6}$ or $x = 1 - \sqrt{6}$. Dividing everything by 3 (and setting the equation equal to 0) gives you $x^2 - 2x - 5 = 0$; you need a leading coefficient of 1 to complete the square, and the simpler equation will yield the same solutions in the quadratic formula.

Completing the Square

$$x^2 - 2x = 5$$

$$x^2 - 2x + 1 = 5 + 1$$

$$\sqrt{(x-1)^2} = \pm\sqrt{6}$$

$$x = 1 \pm \sqrt{6}$$

Quadratic Formula

$$x = \frac{-(-2) \pm \sqrt{(-2)^2 - 4(1)(-5)}}{2(1)}$$

$$x = \frac{2 \pm \sqrt{24}}{2} = \frac{2}{2} \pm \frac{2\sqrt{6}}{2} = 1 \pm \sqrt{6}$$

2. $-2, -1, -\frac{2}{3}, -\frac{1}{2}, -\frac{1}{3}, -\frac{1}{6}, \frac{1}{6}, \frac{1}{3}, \frac{1}{2}, \frac{2}{3}, 1$, and 2. In this function, $a = 6$ (ignore its negative sign) and $c = 2$. Although c has only two factors (1 and 2), a has factors 1, 2, 3, and 6. List every possible fraction you can make with 1 or 2 in the numerator and 1, 2, 3, or 6 in the denominator (and their opposites) to get the final list. You don't have to include $\frac{2}{2}$ and $\frac{2}{6}$ in the list because their simplified forms $\left(1 \text{ and } \frac{1}{3}\right)$ are already there.

3. $-5, \dfrac{-1 + i\sqrt{23}}{6}$, and $\dfrac{-1 - i\sqrt{23}}{6}$. According to Descartes' Rule of Signs, $g(x)$ has no positive roots, but there are either three or one negative root(s). The possible rational roots are $-10, -5, -\frac{10}{3}, -2, -\frac{5}{3}, -1, -\frac{2}{3}, -\frac{1}{3}$ (omitting the positive roots, since this function has none). Only -5 divides evenly using synthetic division, resulting in the quadratic quotient $3x^2 + x + 2$, which is unfactorable. Find its roots using the quadratic formula: $x = \dfrac{^{'}-1 \pm i\sqrt{23}}{6}$.

Chapter 8

1. (a) 125. If $5^3 = x$, then $x = 125$.

   (b) 1. The only exponent that does not change the value of its base, as in the equation $7^x = 7$, is 1.

   (c) $x = -\dfrac{1}{2}$. The exponential version of this equation is $16^x = \dfrac{1}{4}$. Rewrite both sides as a power of 4:

   $$(4^2)^x = 4^{-1}$$

   $$4^{2x} = 4^{-1}$$

   $$2x = -1$$

2. The untransformed graph of $f(x) = \log_5 x$ will pass through the points (1,0) and (5,1), and will treat the y-axis as a vertical asymptote. The graph of $f(x) = -\log_5 (x + 3)$ differs because it has a negative in front of the function (which means you should reflect the graph across the x-axis) and a "+ 3" inside the function (which means you should move all the points of the graph to the *left* three units). In the figure that follows, the original, untransformed graph of $y = \log_5 x$ appears as a dotted curve, and the final graph of $f(x)$ is solid.

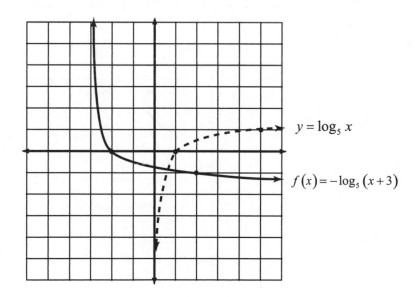

3. 4.248. Both $\dfrac{\log 19}{\log 2} = \dfrac{1.2787536}{0.30102999}$ and $\dfrac{\ln 19}{\ln 2} = \dfrac{2.944438979}{0.69314718}$ equal approximately 4.2479275....

4. (a) $\log 2 - 3\log x - \dfrac{1}{2}\log y$. Rewrite the fraction as a quotient of two logs: $\log 2 - \log\left(x^3\sqrt{y}\right)$. The log on the right contains a product you can rewrite as the sum of two logs: $\log 2 - \log x^3 - \log\sqrt{y}$. (Since $\log\left(x^3\sqrt{y}\right)$ was negative, when it is broken into two logs, they should also be negative.) Write the exponents in front of the logs: $\log 2 - 3\log x - \dfrac{1}{2}\log y$.

    (b) $\ln\dfrac{x-y}{y^3}$. Be careful! You cannot do anything to $\ln(x-y)$ because there is no rule regarding a single log containing subtraction—only *two* logs that are subtracted. Rewrite $3\ln y$ as $\ln y^3$: $\ln(x-y) - \ln y^3$. The difference of two logs can be written as the log of the quotient: $\ln\dfrac{x-y}{y^3}$.

Chapter 9

1. The untransformed graph of $y = 3^x$ passes through the points $(0,1)$ and $(1,3)$, and has a horizontal asymptote at $y = 0$ (the x-axis). It appears in the following diagram as a dotted curve. To get the final graph of $h(x) = 3^{-x} + 1$, reflect $y = 3^x$ across the y-axis and move the entire graph up 1 unit. The graph of $h(x)$ that follows appears as a solid curve.

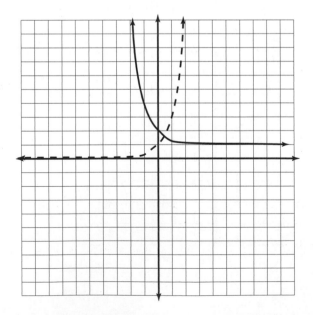

2. (a) $x = 768$. Rewrite the left side as a single $\log_2$ expression: $\log_2 \frac{x}{3} = 8$.

    Exponentiate both sides (using a base of 2), and you get $\frac{x}{3} = 2^8$, or $\frac{x}{3} = 256$. Multiply both sides by 3 for the final answer.

    (b) $x = 1 + \log_8 5$. Take $\log_8$ of both sides of the equation: $\log_8 8^{x-1} = \log_8 5$. Simplify the left side of the equation, and you'll get $x - 1 = \log_8 5$. The most accurate and precise way to write the answer is $x = 1 + \log_8 5$, which (according to the change of base formula) approximately equals 1.774.

3. 36,291 colonies. Since 32 hours pass between 8 A.M. Monday and 4 P.M. Tuesday, set $t = 32$, $N = 9$, and $F = 113$. Plug these into the formula $F = Ne^{kt}$ to calculate k:

    $$113 = 9e^{k \cdot 32}$$

    $$\frac{113}{9} = e^{32k}$$

    $$k = \frac{\ln \frac{113}{9}}{32} \approx 0.079067601293$$

    Therefore, exactly t hours after the scientist spotted 9 colonies, the total colonies will have grown to F, where $F = 9e^{0.079067601293(t)}$. When she leaves at 5 P.M. on Friday, 105 hours will have passed since 8 A.M. Monday, so $F = 9e^{0.079067601293(105)} = 36{,}290.915$, which is approximately 36,291 colonies.

Chapter 10

1. (a) $-\frac{5\pi}{6}$. Multiply by $\frac{\pi}{180}$ and simplify: $\frac{-150}{1} \cdot \frac{\pi}{180} = -\frac{150\pi}{180} = -\frac{5\pi}{6}$.

    (b) 240°. Multiply by $\frac{180}{\pi}$ and simplify: $\frac{4\pi}{3} \cdot \frac{180}{\pi} = \frac{720\pi}{3\pi} = 240°$.

    (c) $\frac{900}{\pi}°$ or 286.479°. Even though this has no π in it, you still multiply by $\frac{180}{\pi}$:

    $$\frac{5}{1} \cdot \frac{180}{\pi} = \frac{900}{\pi}°.$$

2. (a) −540° and 180° are possible answers. Keep adding 360° to −900° until you get a positive number; any negative number you get along the way is a valid negative coterminal:

    $$-900° + 360° = -540° + 360° = -180° + 360° = 180°$$

(b) $\frac{11\pi}{4}$ and $-\frac{5\pi}{4}$ are possible answers. Add and subtract $\frac{8\pi}{4}$ (that's just 2π written with a matching denominator):

$$\frac{3\pi}{4} + \frac{8\pi}{4} = \frac{11\pi}{4} \quad \text{and} \quad \frac{3\pi}{4} - \frac{8\pi}{4} = -\frac{5\pi}{4}$$

3. 1.369. The sides with length 6 and x are, respectively, the opposite and adjacent sides to the acute angle $\frac{3\pi}{7}$, so use tangent: $\tan\frac{3\pi}{7} = \frac{6}{x}$. Multiply both sides by x, and then divide both sides by $\tan\frac{3\pi}{7}$ to get the answer: $x = \dfrac{6}{\tan\frac{3\pi}{7}} \approx 1.3694608$.

Make sure your calculator is set to radians mode.

4. $\sin\frac{21\pi}{4} = -\frac{\sqrt{2}}{2}$. If you subtract $2\pi\left(\frac{8\pi}{4}\right)$ from $\frac{21\pi}{4}$ twice, you'll get a coterminal angle on the unit circle: $\frac{5\pi}{4}$. Therefore, $\frac{21\pi}{4}$ and $\frac{5\pi}{4}$ have equal cosine and sine values based on the unit circle intersection point $\left(-\frac{\sqrt{2}}{2}, -\frac{\sqrt{2}}{2}\right)$.

Chapter 11

1. Period = 7; amplitude = 2.5. Choose corresponding points on consecutive repetitions of the graph, such as $(-6,-2)$ and $(1,-2)$; the period is $|-6-1| = 7$. To calculate the amplitude, choose one instance of the highest and lowest heights reached by the function, such as $(-2,3)$ and $(1,-2)$. Take the absolute value of the difference of the y's, $|3-(-2)| = 5$, and divide by 2.

2. Three transformations change $\sin x$ into $j(x)$: A period change (the graph of $j(x)$ will be twice as squished toward the y-axis, with a period of $\frac{2\pi}{2} = \pi$), an amplitude change (the graph will stretch from -4 to 4 instead of -1 to 1 before you move it), and a vertical shift of the entire graph down 1 unit. The graph of $j(x)$ appears as a solid curve in the figure that follows; the dotted curve is the untransformed graph of $y = \sin x$.

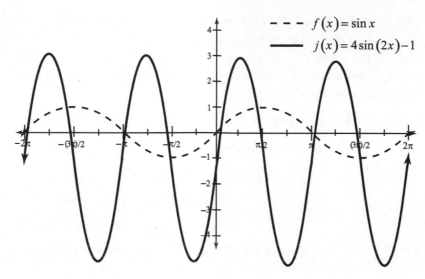

3. From the unit circle, you get $\cos\dfrac{5\pi}{6} = -\dfrac{\sqrt{3}}{2}$ and $\sin\dfrac{5\pi}{6} = \dfrac{1}{2}$, which means:

$$\tan\dfrac{5\pi}{6} = \dfrac{\dfrac{1}{2}}{-\dfrac{\sqrt{3}}{2}}$$

Multiply the numerator and denominator by 2 to simplify: $\tan\dfrac{5\pi}{6} = -\dfrac{1}{\sqrt{3}}$ or rationalize it to get $\tan\dfrac{5\pi}{6} = -\dfrac{\sqrt{3}}{3}$. Take reciprocals to get the other trig functions: $\sec\dfrac{5\pi}{6} = -\dfrac{2}{\sqrt{3}}$ or $-\dfrac{2\sqrt{3}}{3}$, $\csc\dfrac{5\pi}{6} = 2$, and $\cot\theta = -\sqrt{3}$.

Chapter 12

1. $\dfrac{1+\cos^2\theta}{\sin^2\theta}$. The expression is the difference of perfect squares, so factoring gives you $(\csc^2\theta + \cot^2\theta)(\csc^2\theta - \cot^2\theta)$. Massage Baby by subtracting $\cot^2\theta$ from both sides of its identity ($1 = \csc^2\theta - \cot^2\theta$). You can replace $\csc^2\theta - \cot^2\theta$ in the problem with 1: $(\csc^2\theta + \cot^2\theta)(1)$. Rewrite what's left using sine and cosine $\left(\dfrac{1}{\sin^2\theta} + \dfrac{\cos^2\theta}{\sin^2\theta}\right)$ and add, using common denominators. You cannot simplify the numerator further because Mama cannot be massaged to generate $1 + \cos^2\theta$.

2. The right side has a cofunction and a sign identity, so rewrite those: $\cos x - \cos^3 x = \sin^2 x \cos x$. The left side has a greatest common factor of $\cos x$; pull it out: $\cos x(1 - \cos^2 x) = \sin^2 x \cos x$. According to Mama, you can rewrite $(1 - \cos^2 x)$ as $\sin^2 x$, so do that: $\cos x(\sin^2 x) = \sin^2 x \cos x$. Both sides of the equation are equal, so you're done.

3. Rewrite the left side using the sine double-angle formula and the sine sum formula:

$$2\sin\theta\cos\theta\left(\sin\theta\cos\frac{3\pi}{2} + \cos\theta\sin\frac{3\pi}{2}\right) = -2\cos^2\theta\sin\theta$$

$$2\sin\theta\cos\theta\left(\sin\theta\cdot 0 + \cos\theta\cdot -1\right) = -2\cos^2\theta\sin\theta$$

$$2\sin\theta\cos\theta\left(-\cos\theta\right) = -2\cos^2\theta\sin\theta$$

$$-2\sin\theta\cos^2\theta = -2\cos^2\theta\sin\theta$$

Those two sides are equal, according to the commutative property—the order of multiplication doesn't matter.

Chapter 13

1. $-\frac{\pi}{4}$. If an angle has a cosecant of $-\sqrt{2}$, its sine is the reciprocal: $-\frac{1}{\sqrt{2}}$, which equals $-\frac{\sqrt{2}}{2}$ once rationalized. In other words, $\text{arccsc}-\sqrt{2}$ and $\text{arcsin}-\frac{\sqrt{2}}{2}$ have the same value. Two angles have a sine of $-\frac{\sqrt{2}}{2}$: $\theta = \frac{5\pi}{4}$ and $\theta = \frac{7\pi}{4}$, but neither fits the restricted range of $\left[-\frac{\pi}{2},\frac{\pi}{2}\right]$. Throw out the answer $\theta = \frac{5\pi}{4}$ (the arcsine must output angles in quadrants I and IV only), and calculate a negative coterminal angle for $\theta = \frac{7\pi}{4}$: $\frac{7\pi}{4} - 2\pi = \frac{7\pi}{4} - \frac{8\pi}{4} = -\frac{\pi}{4}$.

2. (a) $\theta = 0$. Only $\cos 0 = 1$ on the interval $[0,2\pi)$. Notice that 2π, which also has a cosine value of 1, is not included on the interval because of the parenthesis.

   (b) $\theta = 0$. The restricted range for arccosine is $[0,\pi]$, and the exact solution fits.

   (c) $\theta = 0 + 2k\pi$. All the angles coterminal to 0 (including 2π) are solutions.

3. $\theta = \frac{\pi}{2} + 2k\pi, \frac{3\pi}{2} + 2k\pi$. Distribute the 2 and then isolate $\cot\theta$:

$$2\cot\theta + 6 = 6$$

$$2\cot\theta = 0$$

$$\cot\theta = 0$$

This means that $\theta = \text{arccot } 0$. The cotangent function equals 0 when its numerator (cosine) equals 0 on the unit circle: $\theta = \frac{\pi}{2}$ and $\theta = \frac{3\pi}{2}$. The general solution is both unit circle values with "$+ 2k\pi$" tacked on to each.

4. $\theta = 0, \frac{\pi}{2}, \pi, \frac{7\pi}{6}$, and $\frac{11\pi}{6}$. Subtract $\sin\theta$ from both sides (to set the equation equal to 0), and factor out the greatest common factor, $\sin\theta$: $\sin\theta(2\sin^2\theta - \sin\theta - 1) = 0$. Factor the quadratic in parentheses to get $\sin\theta(2\sin\theta + 1)(\sin\theta - 1) = 0$. Set each factor to 0 and solve:

$\sin\theta = 0$	$\sin\theta - 1 = 0$	$2\sin\theta + 1 = 0$
$\theta = \arcsin 0$	$\theta = \arcsin 1$	$\theta = \arcsin\left(-\frac{1}{2}\right)$
$\theta = 0, \pi$	$\theta = \frac{\pi}{2}$	$\theta = \frac{7\pi}{6}, \frac{11\pi}{6}$

5. $\theta = \frac{\pi}{4}, \frac{3\pi}{4}, \frac{5\pi}{4}, \frac{7\pi}{4}$. Both functions are squared, so you can use Papa to replace either one; I recommend replacing $\sec^2\theta$ with $1 + \tan^2\theta$: $2\tan^2\theta = 1 + \tan^2\theta$. Subtract $\tan^2\theta$ from both sides: $\tan^2\theta = 1$. Now square root both sides to get $\tan\theta = \pm 1$. (The tangent of all four unit circle angles with a denominator of 4 is either 1 or -1.)

6. $\theta = 0, \frac{\pi}{2}$. Subtract $\sin\theta$ from both sides and square them: $(\cos\theta)^2 = (1 - \sin\theta)^2$. Once you square the binomial, you'll get $\cos^2\theta = 1 - 2\sin\theta + \sin^2\theta$. Replace $\cos^2\theta$ with the massaged Mama expression $1 - \sin^2\theta$ so that everything's in terms of sine:

$$1 - \sin^2\theta = 1 - 2\sin\theta + \sin^2\theta$$
$$0 = -2\sin\theta + 2\sin^2\theta$$
$$0 = 2\sin\theta\left(-1 + \sin\theta\right)$$

Set both factors equal to 0; you'll get $\sin\theta = 0$ (whose exact solution is $\theta = 0$) and $\sin\theta = 1$ (whose exact solution is $\theta = \frac{\pi}{2}$). Both solutions work if you test them in the original equation.

7. $\theta = \frac{\pi}{3} + 2k\pi, \frac{2\pi}{3} + 2k\pi, \frac{4\pi}{3} + 2k\pi, \frac{5\pi}{3} + 2k\pi$. Isolate $\cos 2\theta$ by subtracting 1 from and dividing by 2 on both sides: $\cos 2\theta = -\frac{1}{2}$. Use the arccosine function to eliminate the cosine: $2\theta = \frac{2\pi}{3}, \frac{4\pi}{3}$. Because θ's coefficient is 2, you should list twice as many solutions using coterminal angles: $2\theta = \frac{2\pi}{3}, \frac{4\pi}{3}, \frac{8\pi}{3}, \frac{10\pi}{3}$. To finish, multiply everything by $\frac{1}{2}$, simplify, and tack "$+ 2k\pi$" onto each.

Chapter 14

1. $\frac{\pi}{7}$. Since $\theta = \frac{13\pi}{7}$ is a fourth-quadrant angle, you should draw the fourth-quadrant piece of the bowtie using its terminal side, as demonstrated by the figure that follows:

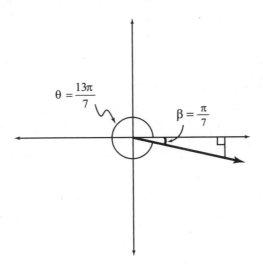

The reference angle (β) will be the small angle sandwiched between θ and the x-axis: $2\pi - \frac{13\pi}{7} = \frac{14\pi}{7} - \frac{13\pi}{7} = \frac{\pi}{7}$.

2. $\sin\alpha = -\frac{5}{\sqrt{89}}$ or $-\frac{5\sqrt{89}}{89}$. If $\sec\alpha = \frac{\sqrt{89}}{8}$, then $\cos\alpha = \frac{8}{\sqrt{89}}$. The only quadrant in which the cosine and secant are positive but the cosecant is negative ($\csc\alpha < 0$) is quadrant IV. Draw the fourth-quadrant piece of the bowtie and label the sides. Since the cosine is the adjacent divided by the hypotenuse, the adjacent side is 8 and the hypotenuse is $\sqrt{89}$. (Both are positive because the horizontal side travels *right* from the origin, not left, and the hypotenuse is never negative.) Use the Pythagorean Theorem to get the remaining side length of 5; it should be negative because you travel *down* from the origin, not up. You'll end up with the following diagram:

Sine equals the opposite divided by the hypotenuse, so $\sin\alpha = -\dfrac{5}{\sqrt{89}}$ or $-\dfrac{5\sqrt{89}}{89}$.

3. 7.112. You're given two sides and an included angle (SAS), so use the Law of Cosines. If you set $a = 5$ and $b = 6$, then $C = 80°$ because it is the angle opposite the side c, whose length is not given:

$$c^2 = a^2 + b^2 - 2ab\cos C$$
$$c^2 = 5^2 + 6^2 - 2(5)(6)\cos 80°$$
$$c^2 = 61 - 60(0.173648178)$$
$$\sqrt{c^2} = \sqrt{50.58110932}$$
$$c \approx 7.112$$

4. $A = 135.951°$, $B = 18.717°$, and $C = 25.332°$. You're given a SSS triangle, which means you have to start with the Law of Cosines (although the Law of Sines will come into play later). You should use the Law of Cosines to determine a triangle's largest angle (in case, it is obtuse), so calculate A, the angle opposite the longest side:

$$a^2 = b^2 + c^2 - 2bc\cos A$$
$$(6.5)^2 = 3^2 + 4^2 - 2(3)(4)\cos A$$
$$42.25 = 25 - 24\cos A$$
$$-0.71875 = \cos A$$
$$135.951374326° \approx A$$

Now that you know the measures of one angle/side pair, (a and A), use the Law of Sines to calculate another missing angle:

$$\frac{\sin(135.951374326)}{6.5} = \frac{\sin B}{3}$$

$$3(0.695268608165) = 6.5(\sin B)$$

$$0.32089320377 \approx \sin B$$

$$18.716950657° \approx B$$

Subtract A and B from $180°$ to get angle C.

5. 8.999. The problem describes a SAS triangle, so use the area formula: $A = \frac{1}{2}ab \cdot \sin C$:

$$\text{Area} = \frac{1}{2}(4)(7) \cdot \sin 40°$$

$$\approx 14(0.6427876097)$$

$$\approx 8.999$$

6. $6\sqrt{95}$ in^2. Begin by calculating s: $s = \dfrac{9+13+16}{2} = 19$. Now plug $a = 9$, $b = 13$, and $c = 16$ into Heron's Area Formula:

$$\text{Area} = \sqrt{19(19-9)(19-13)(19-16)}$$

$$= \sqrt{19(10)(6)(3)}$$

$$= 6\sqrt{95} \text{ in}^2 \approx 58.481 \text{ in}^2$$

Chapter 15

1. (a) Hyperbola. The coefficients of x^2 (3) and y^2 (–1) are unequal and have opposite signs.

   (b) Parabola. There is no x^2 term in the equation. If either squared term is missing, the conic is automatically a parabola.

2. Vertex = (1,3); focus = $\left(1, \dfrac{49}{16}\right)$. Move the constant to the left side of the equation, and factor out the x^2 coefficient: $y - 7 = 4(x^2 - 2x)$. Completing the square means adding 1 inside the parentheses and adding $4 \cdot 1 = 4$ to the left side of the equation:

$y - 7 + 4 = 4(x^2 - 2x + 1)$. Now factor the right side to get $y - 3 = 4(x - 1)^2$, and solve for y to put the answer in standard form: $y = 4(x - 1)^2 + 3$. This means $b = 1$, $k = 3$, and $a = 4$.

To figure out the focus, you need to know two things: the direction the parabola points (up, since a is positive) and c. Since $a = \dfrac{1}{4c}$ and you already know what a is, plug it in:

$$4 = \frac{1}{4c}$$
$$16c = 1$$
$$c = \frac{1}{16}$$

So, the focus will be exactly $\dfrac{1}{16}$ of a unit *above* the vertex. (If the parabola opened downward, the focus would be *below* the vertex.) You already figured out b and k, so you know the vertex is $(1,3)$; add c to its y-value to get the focus: $\left(1, 3\dfrac{1}{16}\right)$ or $\left(1, \dfrac{49}{16}\right)$.

3. $x = (y - 3)^2 - 1$; vertex $= (-1,3)$; focus $= \left(-\dfrac{3}{4}, 3\right)$; directrix is $x = -\dfrac{5}{4}$. Subtract 8 from both sides and then complete the square, just like you did when x and y were reversed in Problem 2: $x - 8 + 9 = y^2 - 6y + 9$. Simplify the left side, and factor the right side to get $x + 1 = (y - 3)^2$. Solve for x, and you've got standard form: $x = (y - 3)^2 - 1$. This means $a = 1$ (there is no other constant in front of the parentheses), $b = -1$, and $k = 3$. The vertex is $(b,k) = (-1,3)$; use a to calculate c:

$$1 = \frac{1}{4c}$$
$$4c = 1$$
$$c = \frac{1}{4}$$

Since a is positive, the parabola points right. The focus is $\dfrac{1}{4}$ units right of the vertex, $\left(-\dfrac{3}{4}, 3\right)$, and the directrix is $\dfrac{1}{4}$ units in the opposite direction: $x = -\dfrac{5}{4}$.

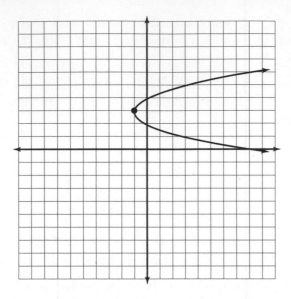

4. $(x + 3)^2 + (y - 5) = 4$. Subtract 30 from both sides to move it across the equals sign, and regroup the left side so that the x and y terms are separated: $(x^2 + 6x) + (y^2 - 10y) = -30$. Complete the square for the x and y quadratics separately: $(x^2 + 6x + 9) + (y^2 - 10y + 25) = -30 + 9 + 25$. Factor and simplify to get standard form: $(x + 3)^2 + (y - 5)^2 = 4$. Since $h = -3$, $k = 5$, and $r^2 = 4$, the center of the circle is $(-3,5)$ and the radius is $\sqrt{4} = 2$, which results in the following graph:

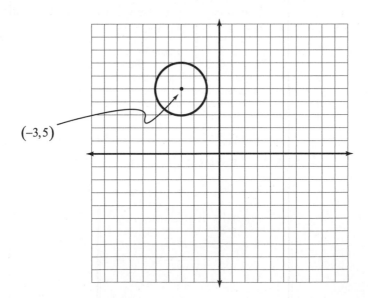

$(-3,5)$

Chapter 16

1. $\dfrac{(x-2)^2}{4} + \dfrac{(y+5)^2}{1} = 1$; major axis length = 4. Rearrange the equation so the x and y terms are grouped, factor 4 out of the y terms, and move 100 to the right side of the equation: $(x^2 - 4x) + 4(y^2 + 10y) = -100$. Complete the square for x and y: $(x^2 - 4x + 4) + 4(y^2 + 10y + 25) = -100 + 4 + 4(25)$. Factor and simplify: $(x - 2)^2 + 4(y + 5)^2 = 4$. Divide everything by 4 to eliminate the constant on the right side of the equation: $\dfrac{(x-2)^2}{4} + \dfrac{(y+5)^2}{1} = 1$. This means $h = 2$, $k = -5$, $a = 2$, and $b = 1$. The major axis is horizontal and has the length $2a = 4$. The graph is pictured in the following diagram:

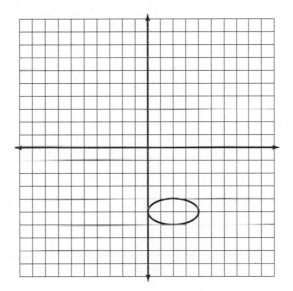

2. $e \approx 0.894$. When you put this ellipse into standard form, it will have some ugly fractions, but don't panic. Complete the square for y to get started:

$$5x^2 + (y^2 - 3y) = -1$$

$$5x^2 + \left(y^2 - 3y + \frac{9}{4}\right) = -1 + \frac{9}{4}$$

$$5x^2 + \left(y - \frac{3}{2}\right)^2 = \frac{5}{4}$$

Divide everything by $\frac{5}{4}$ to get standard form. Notice that $\frac{5}{5/4}$ can be simplified by dividing the top and bottom by 5, which results in $\frac{1}{1/4}$:

$$\frac{x^2}{1/4} + \frac{\left(y - \frac{3}{2}\right)^2}{5/4} = 1$$

This means $h = 0$ (you didn't have to complete the square for x), $k = \frac{3}{2}$, $a^2 = \frac{5}{4}$, and $b^2 = \frac{1}{4}$. Calculate c using the formula $c = \sqrt{a^2 - b^2}$:

$$c = \sqrt{\frac{5}{4} - \frac{1}{4}} = \sqrt{\frac{4}{4}} = 1$$

Finally, divide c by a to get the eccentricity:

$$e = \frac{1}{\sqrt{5/4}} \approx 0.894$$

3. $\frac{(x+2)^2}{16} - \frac{(y+1)^2}{25} = 1$. Group the x and y terms, and move the constant to the other side of the equation: $(25x^2 + 100x) + (-16y^2 - 32y) = 316$. Factor 25 out of the x group and -16 out of the y group: $25(x^2 + 4x) - 16(y^2 + 2y) = 316$. Complete the square for both sets of parentheses: $25(x^2 + 4x + 4) - 16(y^2 + 2y + 1) = 316 + 100 - 16$. (Don't forget to multiply the new constants by the number in front of each one's parentheses when balancing the right side of the equation; add 100 and subtract 16, rather than adding 4 and subtracting 1.)

Factor and simplify: $25(x + 2)^2 - 16(y + 1)^2 = 400$. Divide everything by 400 and reduce the fractions: $\frac{(x+2)^2}{16} - \frac{(y+1)^2}{25} = 1$. This hyperbola is centered at $(-2,-1)$, has a horizontal transverse axis of length 8 (since $a = 4$), and has a vertical conjugate axis of length 10 (since $b = 5$). Its graph will look like the figure on the next page:

Chapter 17

1. $\begin{bmatrix} -10 & -8 \\ 37 & 53 \end{bmatrix}$. Multiply the matrices by their scalars:

$$\begin{bmatrix} -2 \cdot 1 & -2 \cdot 6 \\ -2(-4) & -2 \cdot 3 \end{bmatrix} + \begin{bmatrix} 5(-4) & 5 \cdot 0 \\ 5 \cdot 9 & 5 \cdot 7 \end{bmatrix} + \begin{bmatrix} -4(-3) & -4(-1) \\ -4(4) & -4(-6) \end{bmatrix}$$

Add all corresponding entries:

$$\begin{bmatrix} -2-20+12 & -12+0+4 \\ 8+45-16 & -6+35+24 \end{bmatrix} = \begin{bmatrix} -10 & -8 \\ 37 & 53 \end{bmatrix}$$

2. $\begin{bmatrix} 37 & 34 \\ -56 & -90 \\ -11 & -28 \end{bmatrix}$. The product matrix has three rows and two columns:

$$\begin{bmatrix} 5 & -3 & -2 \\ -10 & 4 & 8 \\ -1 & 1 & 6 \end{bmatrix} \cdot \begin{bmatrix} 2 & 7 \\ -9 & 3 \\ 0 & -4 \end{bmatrix} = \begin{bmatrix} 5(2)+(-3)(-9)+(-2)(0) & 5(7)+(-3)(3)+(-2)(-4) \\ -10(2)+4(-9)+8(0) & -10(7)+4(3)+8(-4) \\ (-1)(2)+1(-9)+6(0) & (-1)7+1(3)+6(-4) \end{bmatrix}$$

3. 21. Be sure to subtract in the right order: 8(2) − (5)(−1) = 16 + 5 = 21.

4. 132. Create the matrix $\begin{bmatrix} 3 & -2 & 5 & 3 & -2 \\ 1 & 9 & 7 & 1 & 9 \\ 0 & -1 & 4 & 0 & -1 \end{bmatrix}$. Multiplying the diagonals correctly

gives you this expression:

$$3(9)(4) + (-2)(7)(0) + 5(1)(-1) - 0(9)(5) - (-1)(7)(3) - (4)(1)(-2)$$

After multiplying, you get $108 + 0 - 5 - 0 + 21 + 8$, which equals 132.

5. $M_{23} = 12$; $C_{23} = -12$. The element -2 occurs in the second row and the third column, meaning $-2 = b_{23}$. (In other words, $i = 2$ and $j = 3$.) The minor, M_{23}, is the determinant of the matrix without the row and column containing -2:

$$M_{23} = \begin{vmatrix} 1 & -3 \\ 4 & 0 \end{vmatrix} = 0 - (-12) = 12$$

The cofactor, C_{23} equals $(-1)^{i+j}$ times M_{23}:

$$(-1)^{2+3}(12) = (-1)^5(12) = -12$$

6. 132. You should expand either row 3 or column 1, since both contain $a_{31} = 0$. I will expand column 1:

$$3 \cdot (-1)^{1+1} \cdot \begin{vmatrix} 9 & 7 \\ -1 & 4 \end{vmatrix} + 1 \cdot (-1)^{2+1} \cdot \begin{vmatrix} -2 & 5 \\ -1 & 4 \end{vmatrix} + 0 \cdot (-1)^{3+1} \cdot \begin{vmatrix} -2 & 5 \\ 9 & 7 \end{vmatrix}$$
$$= 3(36+7) - (-8+5)$$
$$= 3(43) + 3$$
$$= 132$$

7. $\left(-\frac{1}{5}, \frac{2}{25} \right)$. Create matrices C, X, and Y:

$$C = \begin{bmatrix} -6 & 10 \\ 1 & 15 \end{bmatrix} \qquad X = \begin{bmatrix} 2 & 10 \\ 1 & 15 \end{bmatrix} \qquad Y = \begin{bmatrix} -6 & 2 \\ 1 & 1 \end{bmatrix}$$

Calculate their determinants:

$$|C| = -90 - 10 = -100 \qquad |X| = 30 - 10 = 20 \qquad |Y| = -6 - 2 = -8$$

Divide $|X|$ and $|Y|$ by $|C|$ to get x and y, respectively:

$$x = \frac{|X|}{|C|} \qquad y = \frac{|Y|}{|C|}$$

$$x = \frac{20}{-100} \qquad y = \frac{-8}{-100}$$

$$x = -\frac{1}{5} \qquad y = \frac{2}{25}$$

Chapter 18

1. (a) $\begin{bmatrix} -2 & -3 & 1 & \vdots & 0 \\ -30 & -54 & -12 & \vdots & 36 \\ 4 & -1 & -1 & \vdots & 2 \end{bmatrix}$. Multiply each element of the second row by -6.

(b) $\begin{bmatrix} 1 & -8 & \vdots & 6 \\ 0 & 19 & \vdots & -20 \end{bmatrix}$. Multiply the first row by -3, and add it to the second row:

$$B = \begin{bmatrix} 1 & -8 & \vdots & 6 \\ -3(1)+3 & -3(-8)+(-5) & \vdots & -3(6)+(-2) \end{bmatrix} = \begin{bmatrix} 1 & -8 & \vdots & 6 \\ 0 & 19 & \vdots & -20 \end{bmatrix}$$

2. $\begin{bmatrix} 1 & -\frac{3}{4} & \vdots & \frac{3}{4} \\ 0 & 1 & \vdots & -1 \end{bmatrix}$. Multiply R_1 by $-\frac{1}{4}$ to change a_{11} into 1:

$$\begin{bmatrix} \left(-\frac{1}{4}\right)(-4) & \left(-\frac{1}{4}\right)(3) & \vdots & \left(-\frac{1}{4}\right)(-3) \\ 2 & -1 & \vdots & 1 \end{bmatrix} = \begin{bmatrix} 1 & -\frac{3}{4} & \vdots & \frac{3}{4} \\ 2 & -1 & \vdots & 1 \end{bmatrix}$$

Now apply the row operation $-2 \cdot R_1 + R_2 \to R_2$ to make $a_{21} = 0$:

$$\begin{bmatrix} 1 & -\frac{3}{4} & \vdots & \frac{3}{4} \\ -2(1)+2 & -2\left(-\frac{3}{4}\right)-1 & \vdots & -2\left(\frac{3}{4}\right)+1 \end{bmatrix} = \begin{bmatrix} 1 & -\frac{3}{4} & \vdots & \frac{3}{4} \\ 0 & \frac{1}{2} & \vdots & -\frac{1}{2} \end{bmatrix}$$

Change a_{22} to 1 $(2R_2 \rightarrow R_2)$:

$$\begin{bmatrix} 1 & -\dfrac{3}{4} & \vdots & \dfrac{3}{4} \\ 2(0) & 2\left(\dfrac{1}{2}\right) & \vdots & 2\left(-\dfrac{1}{2}\right) \end{bmatrix} = \begin{bmatrix} 1 & -\dfrac{3}{4} & \vdots & \dfrac{3}{4} \\ 0 & 1 & \vdots & -1 \end{bmatrix}$$

You're finished because the diagonal is complete.

3. $\begin{bmatrix} 1 & 0 & \vdots & 0 \\ 0 & 1 & \vdots & -1 \end{bmatrix}$. Change $a_{12} = -\dfrac{3}{4}$ into 0 via the row operation $\dfrac{3}{4}R_2 + R_1 \rightarrow R_1$:

$$\begin{bmatrix} \dfrac{3}{4}(0)+1 & \dfrac{3}{4}(1)+\left(-\dfrac{3}{4}\right) & \vdots & \dfrac{3}{4}(-1)+\dfrac{3}{4} \\ 0 & 1 & \vdots & -1 \end{bmatrix} = \begin{bmatrix} 1 & 0 & \vdots & 0 \\ 0 & 1 & \vdots & -1 \end{bmatrix}$$

4. $\begin{bmatrix} -3 & 0 \\ 3 & 2 \end{bmatrix}$. Augment the matrix with a 2×2 identity matrix and work toward reduced row-echelon form on the left matrix. Start with the row operations $-3R_1 \rightarrow R_1$ and $-\dfrac{1}{2}R_1 + R_2 \rightarrow R_2$:

$$\begin{bmatrix} (-3)\left(-\dfrac{1}{3}\right) & (-3)(0) & \vdots & (-3)(1) & (-3)(0) \\ -\dfrac{1}{2}(1)+\dfrac{1}{2} & -\dfrac{1}{2}(0)+\dfrac{1}{2} & \vdots & -\dfrac{1}{2}(-3)+0 & -\dfrac{1}{2}(0)+1 \end{bmatrix} = \begin{bmatrix} 1 & 0 & \vdots & -3 & 0 \\ 0 & \dfrac{1}{2} & \vdots & \dfrac{3}{2} & 1 \end{bmatrix}$$

Now apply $2R_2 \rightarrow R_2$, and you'll have reduced row-echelon form in the left matrix:

$$\begin{bmatrix} 1 & 0 & \vdots & -3 & 0 \\ 2(0) & 2\left(\dfrac{1}{2}\right) & \vdots & 2\left(\dfrac{3}{2}\right) & 2(1) \end{bmatrix} = \begin{bmatrix} 1 & 0 & \vdots & -3 & 0 \\ 0 & 1 & \vdots & 3 & 2 \end{bmatrix}$$

The 2×2 matrix on the right is the inverse.

5. $\begin{bmatrix} x \\ y \end{bmatrix} = \begin{bmatrix} -1 \\ -\frac{3}{2} \end{bmatrix}$. Create an augmented matrix to calculate the inverse of $\begin{bmatrix} -1 & -2 \\ 0 & 2 \end{bmatrix}$.

Once you apply the row operations $-1 \cdot R_1 \to R_1$, $\frac{1}{2} R_2 \to R_2$, and $-2R_2 + R_1 \to R_1$,

you'll get an inverse matrix of $\begin{bmatrix} -1 & -1 \\ 0 & \frac{1}{2} \end{bmatrix}$. Multiply that on both sides of the equation:

$$\begin{bmatrix} -1 & -1 \\ 0 & \frac{1}{2} \end{bmatrix} \cdot \begin{bmatrix} -1 & -2 \\ 0 & 2 \end{bmatrix} \cdot \begin{bmatrix} x \\ y \end{bmatrix} = \begin{bmatrix} -1 & -1 \\ 0 & \frac{1}{2} \end{bmatrix} \cdot \begin{bmatrix} 4 \\ -3 \end{bmatrix}$$

$$\begin{bmatrix} x \\ y \end{bmatrix} = \begin{bmatrix} -1 \\ -\frac{3}{2} \end{bmatrix}$$

Glossary

acute angle   Measures less than 90° (or $\frac{\pi}{2}$ radians).

additive identity   Zero, since adding 0 to any number a does not change a's value.

adjacent side   The shorter of the two segments that form an acute angle in a right triangle.

amplitude   Half the height of a periodic function, calculated by subtracting the y-values of its highest and lowest points and dividing its absolute value by 2.

angle   Geometric figure created by connecting two rays at their endpoints.

argument (of a logarithmic expression)   In the expression $\log_a x$, x is the argument of the logarithm.

associative property   The grouping of a sum or product does not alter its value, so $(a + b) + c = a + (b + c)$ and $(ab)c = a(bc)$.

asymptote   A line representing unattainable points for the graph of a rational function. It's usually drawn as a dotted line on the coordinate plane but is not technically part of the graph.

augmented matrix   Contains both the coefficient matrix and the column of constants, usually separated from the coefficients by a dotted line.

axiom   *See* property.

axis of symmetry Line that cuts through the middle of a parabola, intersecting at only one point, the vertex.

Baby The trigonometric identity $1 + \cot^2 \theta = \csc^2 \theta$.

base (of an exponential expression) In the expression x^7, the base is x.

base (of a logarithm) In the expression $\log_a x$, a is the base of the logarithm.

branch One of two u-shape pieces that together make up the graph of a hyperbola.

center (of a circle) The fixed point from which all points in a circle are equidistant.

center (of an ellipse) The midpoint of the segment whose endpoints are either the foci or the vertices of the ellipse.

center (of a hyperbola) The midpoint of the segment whose endpoints are either the foci or the vertices of the hyperbola.

change of base formula Helps you calculate logarithmic values that have bases other than 10 and e: $\log_a x = \dfrac{\log x}{\log a}$ or $\log_a x = \dfrac{\ln x}{\ln a}$.

circle Set of points in the coordinate place that are all the same distance (called the radius) from a fixed point (called the center).

closed interval An interval on which both endpoints are included: $[a,b]$.

coefficient The number in a polynomial term; for example, the coefficient of the term $7x^3$ is 7.

cofactor Number (corresponding to matrix element a_{ij}) that results when the minor of a_{ij} is multiplied by $(-1)^{i+j}$.

cofunctions Two functions with the same name (except that one begins with the prefix "co"); sine and cosine are cofunctions, as are secant and cosecant.

common denominators Equal denominators of two or more rational expressions.

common logarithm A logarithm with a base of 10; common logs are usually written without a base, such as "$\log x$" rather than "$\log_{10} x$."

commutative property The order of a sum or product does not alter its value, so $a + b = b + a$, $a \cdot b = b \cdot a$, and *ABBA = rockin'*.

complex number Has the form $a + bi$, where a and b are real numbers. Any imaginary or real numbers are automatically complex as well.

composition of functions   The process of plugging one function into another, written $f(g(x))$ or $(f \circ g)(x)$.

compound inequality   An inequality statement possessing two boundaries that surround a variable expression—$a < x < b$.

conic sections   The collection of four geometric figures (parabolas, circles, ellipses, and hyperbolas) that are cross-sections of a right circular cone when sliced by a plane.

conjugate   The complex number created by changing the middle sign of the given complex number. For example, the conjugate of $a + bi$ is $a - bi$, and vice versa.

conjugate axis   The segment that is perpendicular to the transverse axis at the center of a hyperbola.

constant   A number not multiplied by a variable. Because its value cannot vary, it remains constant, hence the name.

coterminal   Describes angles that have the same terminal side when drawn in standard position.

counting number   *See* natural number.

Cramer's Rule   The quotients of the determinants of a coefficient matrix (once columns are strategically replaced by the constants of the system) provide the solutions to that system.

critical number   A value that causes an expression to equal 0 or become undefined; in a rational expression, critical numbers give either the numerator or the denominator a value of 0.

cube root   A radical expression with an index of 3, such as $\sqrt[3]{y}$.

degree (of a polynomial)   The highest exponent in a polynomial.

degree (unit of angle measurement)   Equivalent to $\frac{1}{360}$ of a circle and denoted using the $°$ symbol.

dependent variable   Variable in a function (usually $f(x)$) whose value varies based upon what you decide to plug in for the independent variable. For example, the value of $f(x)$ in the function $f(x) = x^2$ *depends* upon what you decide to plug in for x.

Descartes' Rule of Signs   The number of sign changes in a function (or that number minus a multiple of 2) corresponds to the number of possible positive roots for the function. A similar number of sign changes in $f(-x)$ gives you the possible number of negative roots.

determinant   Real number defined for any square matrix.

directrix   The fixed line used to define a parabola. Each point on the parabola must be equidistant from the directrix and the focus.

distributive property   You can multiply through a sum or a difference: $a(b + c) = ab + ac$ and $a(b - c) = ab - ac$.

dividend   In the division problem $b\overline{)a}$, a is the dividend.

divisor   In the division problem $b\overline{)a}$, b is the divisor.

domain   The set of possible inputs for a function.

eccentricity   Defined as $\frac{c}{a}$, a value between 0 and 1 that describes the "ovalness" of an ellipse. The closer to 0 it is, the more the ellipse resembles a circle, but the closer to 1 it is, the more pronounced the oval shape of the ellipse.

element   One of the numbers within a matrix, usually written a_{ij}, meaning that it's in row i and column j of matrix A; it's also called an *entry*.

ellipse   Set of points on the coordinate plane such that the sum of the distances from each point to two fixed points (called the foci) is constant.

entry   *See* element.

Euler's number   An irrational constant, labeled e, that is approximately equal to 2.718281828…. It is the base of natural logarithmic and exponential functions.

exact form   Provides only one solution to a trig equation—the solution that falls on the restricted range of the corresponding inverse trig function.

exponent   In the expression x^7, the exponent is 7.

exponential function   Functions that contain a variable exponent, as in $f(x) = 2^x$ and $g(x) = 11^x$.

exponentiation   The process of changing the sides of an equation into exponents of a matching base. It is used to eliminate logarithmic functions from an equation.

factor   If a is a factor of b, the quotient $b \div a$ will not have a remainder, and you can say that a "divides evenly" into b.

focus (of a conic section)   A fixed point that defines the graph of a conic section. It serves a slightly different purpose for each conic, but it is never actually part of the graph.

focus (of an ellipse)   One of two fixed points in the plane that defines an ellipse. The sum of the distances from the foci to each point on an ellipse must be constant.

focus (of a hyperbola)   One of two fixed points in the plane that define a hyperbola. The difference of the distances from the foci to each point on a hyperbola must be constant.

function   Mathematical rule such that any input corresponds to exactly one output.

Gaussian elimination   *See* row-echelon form.

general form   Lists all possible solutions to a trig equation (by taking coterminal angles into account).

greatest common factor   The largest factor that divides evenly into every term of a polynomial.

greatest integer function   Denoted $[\![a]\!]$, it outputs the largest integer that is less than or equal to a.

half-life   The length of time it takes for the mass of an element (usually radioactive) to halve.

half-open interval   An interval that contains exactly one of its endpoints: $[a,b)$ or $(c,d]$ (also called "half-closed").

Heron's Area Formula   Any triangle with sides a, b, and c has the area $\sqrt{s(s-a)(s-b)(s-c)}$, where $s = \dfrac{a+b+c}{2}$.

hyperbola   Set of points on a plane such that the difference of the distances from each point on the graph to two fixed points (called the foci) is constant.

hypotenuse   The longest side of a right triangle. Each acute angle in a right triangle has the hypotenuse as one of its sides.

identity   An equation that is true no matter what angle θ is substituted into it.

identity matrix   Given a square matrix, B, the identity matrix, B_I, has matching order and contains all zeros, except in the diagonal beginning in the upper-left corner and ending in the lower-right corner—those elements are all 1.

identity properties   The value of a real number is unchanged if you add 0 to it or multiply it by 1.

imaginary number   Contains i, which is understood to have a value of $i = \sqrt{-1}$.

independent variable   Variable in a function whose value you can control (usually x). Whichever variable you substitute values into is the independent one.

index (of a radical expression)   In the expression $\sqrt[b]{a}$, b is the index.

initial side   Ray marking the beginning of an angle; usually not discernible from the *terminal side* of an angle, unless the angle is in *standard form*, in which case the initial side lies on the positive x-axis of the coordinate plane.

integer   A number with no obvious fraction or decimal part.

interval notation   A shorthand method for expressing inequality statements consisting of two boundary numbers surrounded by some combination of parentheses or brackets, depending upon whether each boundary is included in the interval.

inverse functions   If $f(g(x)) = g(f(x)) = x$, then $f(x)$ and $g(x)$ are inverse functions.

inverse matrix   The matrix A^{-1}, designed so that $(A)(A^{-1})$ equals the identity matrix.

inverse properties   The sum of a number and its opposite equals 0, and the product of a number and its reciprocal is 1.

irrational number   A number that cannot be expressed as a fraction, whose decimal form neither repeats nor terminates.

Law of Cosines   The sides a, b, and c of an oblique triangle and angle A, which is opposite side a, are related according to this equation: $a^2 = b^2 + c^2 - 2bc \cos A$.

Law of Sines   The sides a, b, and c of an oblique triangle and the angles opposite them (A, B, and C respectively) are related according to this equation:

$$\frac{a}{\sin A} = \frac{b}{\sin B} = \frac{c}{\sin C} \ .$$

leading coefficient   The coefficient of the polynomial term with the highest exponent; for example, the leading coefficient of $4x^3 + x^2 - 3x^5 + 9$ is -3.

leading coefficient test   The end behavior of a function (what direction the function is going as its left and right ends speed off to infinity) can be determined using its leading coefficient and degree.

least common denominator   The smallest possible common denominator for a group of rational expressions.

like terms   Polynomial terms with matching variables, which can then be added to or subtracted from one another.

logarithm   Expression written $\log_a x$ that answers the question "To what power must I raise the base, a, to get x?"; every logarithmic equation $\log_a x = y$ can be rewritten in exponential form: $a^y = x$.

major axis   The segment connecting the vertices of an ellipse, the longer of the two perpendicular segments passing through the ellipse's center.

Mama   The trigonometric identity $\cos^2 \theta + \sin^2 \theta = 1$.

matrix   A rectangular collection of numbers (called *elements* or *entries*), organized in rows and columns and surrounded by brackets.

minor   The determinant of a_{ij}'s matrix, once row i and column j are removed, denoted M_{ij}.

minor axis   The shorter of the two perpendicular segments passing through an ellipse's center.

multiplicative identity   Since $a \cdot 1 = a$ for any real number a, 1 is the multiplicative identity.

natural logarithm   A logarithm with base e, written as "ln" rather than "$\log_e x$."

natural number   A number in the set 1, 2, 3, 4, 5, ...; also called a *counting number*.

oblique   Describes an angle not measuring 90°, or a triangle that does not contain a right angle.

obtuse angle   Measures more than 90° ($\frac{\pi}{2}$ radians), but less than 180° (π radians).

open interval   An interval in which neither endpoint is included: (a, b).

opposite   The product of a number and –1. For instance, the opposite of –7 is $(-1)(-7) = 7$.

opposite side   Of the two smaller sides in a right triangle, this side does not touch the vertex of the angle you're plugging into a trig ratio.

order   Describes the dimensions of a matrix, written $r \times c$ (where r is the number of rows and c is the number of columns).

Papa   The trigonometric identity $1 + \tan^2 \theta = \sec^2 \theta$.

parabola   The set of points on the coordinate plane that are equidistant from a fixed point (called the *focus*) and a fixed line (called the *directrix*).

period   The horizontal distance after which a periodic function repeats itself.

periodic   A function or graph that repeats itself over and over again, like it's in an infinite loop.

point-slope formula   If a line has slope m and contains the point (a,b), then it has equation $y - b = m(x - a)$.

polynomial   The sum (or difference) of little clumps (called *terms*), which are made up of numbers (called *coefficients*) and variables (usually raised to exponents) multiplied together.

power   *See* exponent.

prime   A number or polynomial whose only factors are 1 and the number or polynomial itself; 5 is a prime number (its only factors are 1 and 5), and $x - 3$ is a prime binomial (its only factors are $(x - 3)$ and 1).

property   A mathematical fact that is so basic and fundamental, it is accepted as truth, even though it cannot be verified by rigorous proof (also called an *axiom*).

quadrantal   Describes an angle whose measure is a multiple of 90°; when graphed in standard form, its terminal side lies on the x- or y-axis.

quadratic formula   The solutions to the quadratic equation $ax^2 + bx + c = 0$ are
$$x = \frac{-b \pm \sqrt{b^2 - 4ac}}{2a} \; .$$

radian   An angle that cuts out an arc of a circle whose length is equal to the radius of that circle measures exactly 1 radian, approximately 57.296°.

radical expression   Expression containing a radical sign, such as $\sqrt[x]{b}$.

radicand   In the expression $\sqrt[b]{a}$, a is the radicand.

radius   The distance between the center of a circle and any point on that circle.

range   The set of possible outputs for a function.

rational   Describes something that can be written as a fraction (or a terminating or repeating decimal, if it's a number).

rational root test   Guarantees that any rational roots of the function with the leading coefficient a and the constant c will look like $\pm\frac{y}{x}$, where x is a factor of a and y is a factor of c.

real number   Any number (whether rational or irrational, positive or negative) that can be expressed as a decimal.

reciprocal   Every nonzero real number, x, has a corresponding reciprocal, $\frac{1}{x}$; the reciprocal of a rational number reverses the numerator and denominator of that number (the reciprocal of $-\frac{9}{7}$ is $-\frac{7}{9}$).

reduced row-echelon form   Version of an $n \times n$ matrix containing 1s in the diagonal (a_{11}, a_{22}, ..., a_{nn}) and all 0 elements below and above that diagonal. The process of putting a matrix into reduced row-echelon form is called *Gauss-Jordan elimination*.

reference angle   An acute angle whose trig values match those of an obtuse triangle.

Remainder Theorem   If a function $f(x)$ is divided by the binomial $(x - a)$, the remainder is equal to $f(a)$.

restricted graph   Small segment of a larger (often periodic) graph that passes the horizontal line test and (therefore) has a valid inverse function.

right angle   Measures 90° ($\frac{\pi}{2}$ radians).

root (of a function)   The numbers that make a function equal zero. If c is a root of $g(x)$, then $g(c) = 0$.

row-echelon form   Version of an $m \times n$ matrix containing 1s in the diagonal (a_{11}, a_{22}, ..., a_{mm}) and all 0 elements below that diagonal. The process of putting a matrix into row-echelon form is called *Gaussian elimination*.

row operations   Three things you're allowed to do to the rows of a matrix (rearrange them, multiply them by a constant, and add them to one another) that won't change the solution to the system of equations the matrix represents.

scalar   Number multiplied by elements in a matrix.

singular   Describes a matrix that does not have an inverse.

slope-intercept form   A linear equation solved for y is said to be in slope-intercept form $y = mx + b$, where m is the slope and b is the y-intercept.

specified form   Lists all valid solutions to a trig equation in a specific interval.

square matrix   Matrix containing the same number of rows and columns.

square root   A radical expression with an index of 2; radical signs missing an index have an implied index of 2, such as $\sqrt{7}$.

standard form   Specific format for an equation that ensures the uniformity of solutions. Any correct solution will look exactly the same once it is in standard form. (Different types of equations have different standard forms. For example, a linear equation is in standard form if it looks like $Ax + By = C$, where A and B are integers and $A > 0$.)

standard position (of an angle)   Describes an angle graphed in the coordinate plane such that its vertex lies on the origin and its initial side overlaps the positive x-axis.

straight angle   Measures 180° (π radians).

symmetric property   You can reverse the sides of an equation without affecting its solution; in other words, if $x = y$, $y = x$.

synthetic division   Calculates the quotient of a polynomial dividend and a linear binomial divisor using only the coefficients and constants of the expressions.

term   One "clump" of a polynomial, usually the product of a coefficient and a variable raised to an exponent.

terminal side   The ray at which an angle ends.

transitive property   If $a = b$ and $b = c$, then $a = c$.

transverse axis   Segment whose endpoints are the vertices of a hyperbola. It is perpendicular to the conjugate axis at the hyperbola's center.

unit circle   A circle centered at the origin whose radius is 1; generally used to calculate and memorize the cosine and sine values of common angles.

vertex (of an angle)   The intersection point of the two rays forming an angle.

vertex (of a parabola)   The point on the axis of symmetry that's c units away from both the focus and directrix.

whole number   A number in the set 0, 1, 2, 3, 4, 5, ….

Index

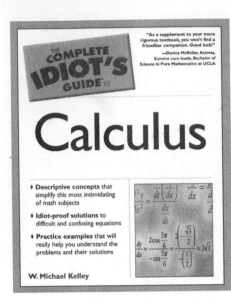